Introduction to Geology

Rob Kanen (Ed) BSc

Introduction to geology.

Bibliography.
ISBN 0 9756723 2 0.

1. Geology. I. Kanen, Rob. II. Mineral Services (Vic.).

551

Table of Contents

Introduction

The first section on rocks is a guide to the terminology and criteria used in the classification of rocks as well as a general introduction to the many different rocktypes identified by geologists. The determinative charts and tables set out the criteria for classifying rocks and use the international guidelines as set out by the International Union of Geological Sciences (IUGS). The diagrams and descriptions use the modal mineralogy observed in a fresh, unweathered sample. Where chemical criteria are used to classify the rock, a standard whole rock chemical analysis is required. In the section on terminology, the terms used in the rock descriptions are defined. There are many additional terms used by geologists; however these are the most widely used terms. Over 300 rock descriptions are included in the section on rocks. Each description follows the same format, consisting of:

Rockname
Group
Family
Occurrence
Texture
Composition
Comments

The second section is a general introduction to the concepts of geological science with chapters on mineralogy, petrology, volcanoes, granites and plate tectonics. The first two chapters on minerals and rocks are from "Textbook of Geology" by Longwell, Knopf and Flint (1934) and "Principles of Geology" by Tyrell (1926). The first section on rocks and all the other chapters on granites, volcanoes and plate tectonics are by the editor and author.

Introduction to Geology

Determinative Tables

Igneous Rocks

The following igneous rock classification diagrams cover most igneous rock types.

Carbonatites - CaO/MgO/FeO+Fe2O3+MnO carbonatites. Contains > 50% primary carbonates.

Charnockites - Quartz/Alkali Feldspar/Plagioclase Charnockites. Contains >5% hypersthene

Melilite Plutonic - Melilite/Olivene/Clinopyroxene melilite plutonic rocks. Contains >10% melilite and <10% feldspathoids

Melilite Volcanic - Melilite/Olivene/Clino-Pyroxene melilite volcanic rocks. Contains >10% melilite and <10% feldspathoids

General Plutonic QAPF - Quartz/Alkali Feldspar/Plagioclase/Foids plutonic rocks. Contains 0-90% mafic minerals.

General Plutonic QAP - Quartz/Alkali Feldspar/Plagioclase plutonic rocks. Contains 0-90% mafic minerals.

General Plutonic FAP - Foid/Alkali Feldspar/Plagioclase plutonic rocks. Contains feldspathoids and 0-90% mafic minerals.

Gabbroic PlagPxOl - Plagioclase/Pyroxene/Olivene mafic rocks. Rocks falling into the gabbroic field of QAP diagram.

Gabbroic PlagOpxCPx - Plagioclase/Orthopyroxene/Clinopyroxene mafic rocks. Rocks falling into the gabbroic field of QAP diagram.

Gabbroic Plag/Px/Hbl - Plagioclase/Pyroxene/Hornblende mafic rocks. Rocks falling into the gabbroic field of QAP diagram.

Ultramafic Ol/OPx/CPx - Olivene/Orthopyroxene/Clinopyroxene ultramafic rocks. Contains >90% mafic minerals.

Ultramafic Ol/Px/Hbl - Olivene/Pyroxene/Hornblende ultramafic rocks. Contains >90% mafic minerals.

Volcanic QAPF - Quartz/Alkali Feldspar/Plagioclase/Foids volcanic rocks. Contains 0-90% mafic minerals.

Volcanic TAS - K+Na vs SiO2 chemical classification for volcanic rocks.
Volcanic TAS - K+Na vs SiO2 chemical classification for volcanic rocks.

Mafic minerals = amphibole + pyroxene + olivene + mica + opaque minerals + epidote + allanite + garnet + melilite + monticellite + primary carbonate + accessory minerals.

Abbreviations used in the tables:

M = mafic minerals

An = Anorthite

Mel = melilite

Q=Quartz

A=Alkali Feldspar

P=Plagioclase

Cpx=Clino-pyroxene

Opx=Orthopyroxene

Px=Pyroxene

Ol=Olivene

Hbl=Hornblende

F=Feldspathoids

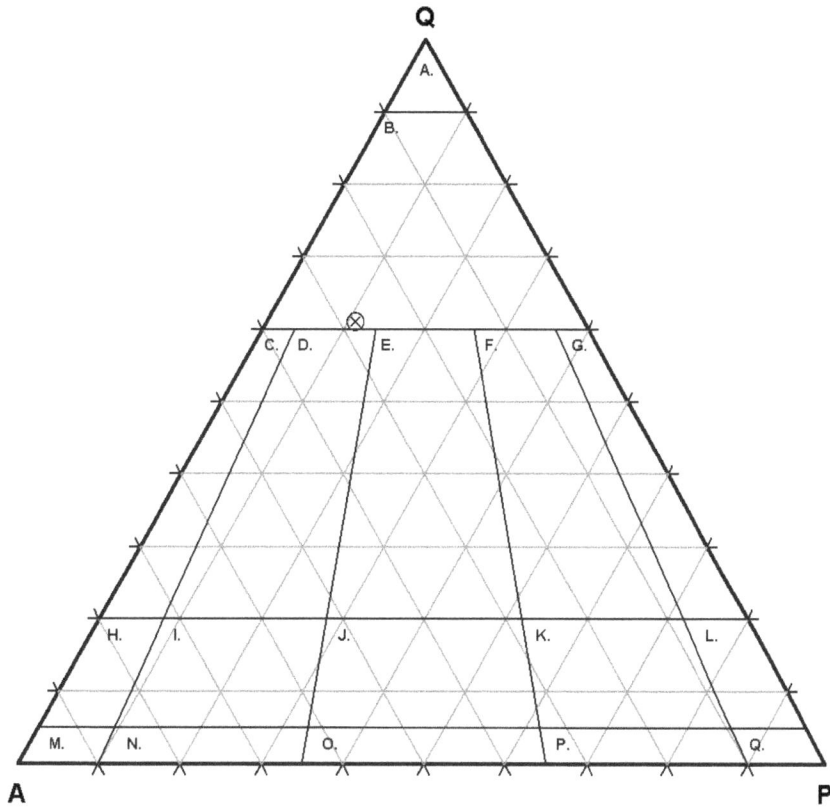

To plot a data point on ternary diagrams, the mineralogy is recalculated into the three end components:

i.e. Original Rock - Quartz 55% Alkali Feldspar 25% Plagioclase 10%

Recalculate to sum to 100% - Quartz = 55/55+25+10 = 61%

Alkali Feldspar = 25/55+25+10 = 28%

Plagioclase = 10/55+25+10 = 11%

Now, each end member can be plotted on the ternary or diamond diagram

Carbonatites

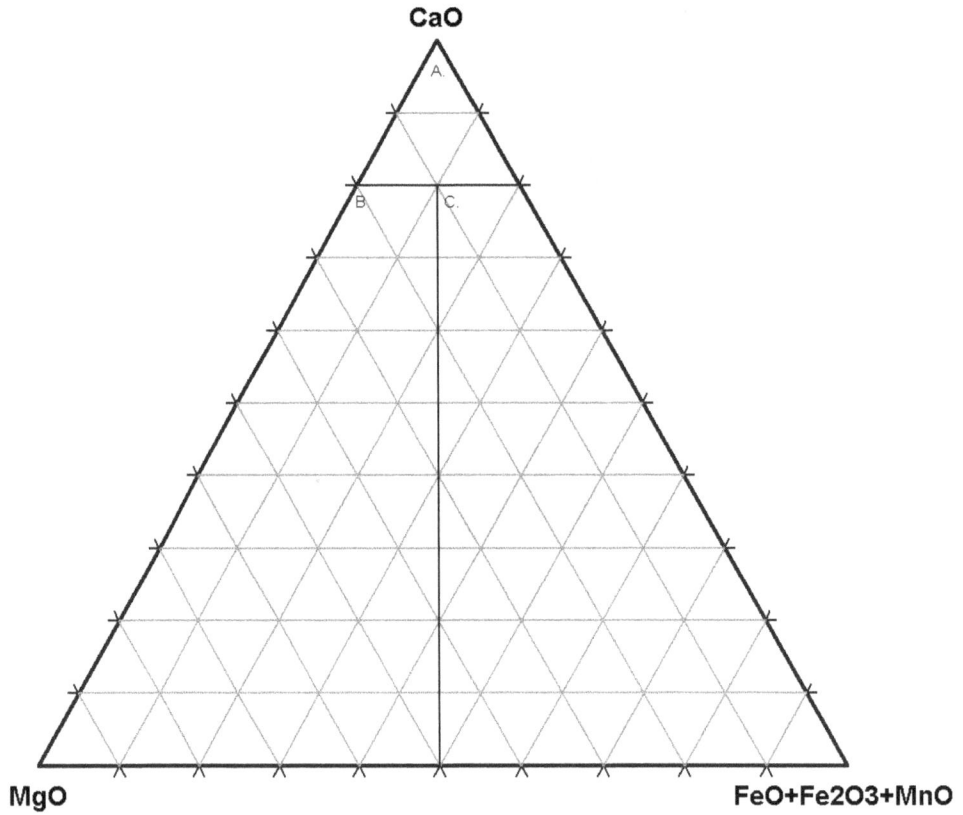

A. Calcio-Carbonatite
B. Magnesiocarbonatite
C. Ferrocarbonatite

Definition: Contains >50% primary carbonates

Charnockites

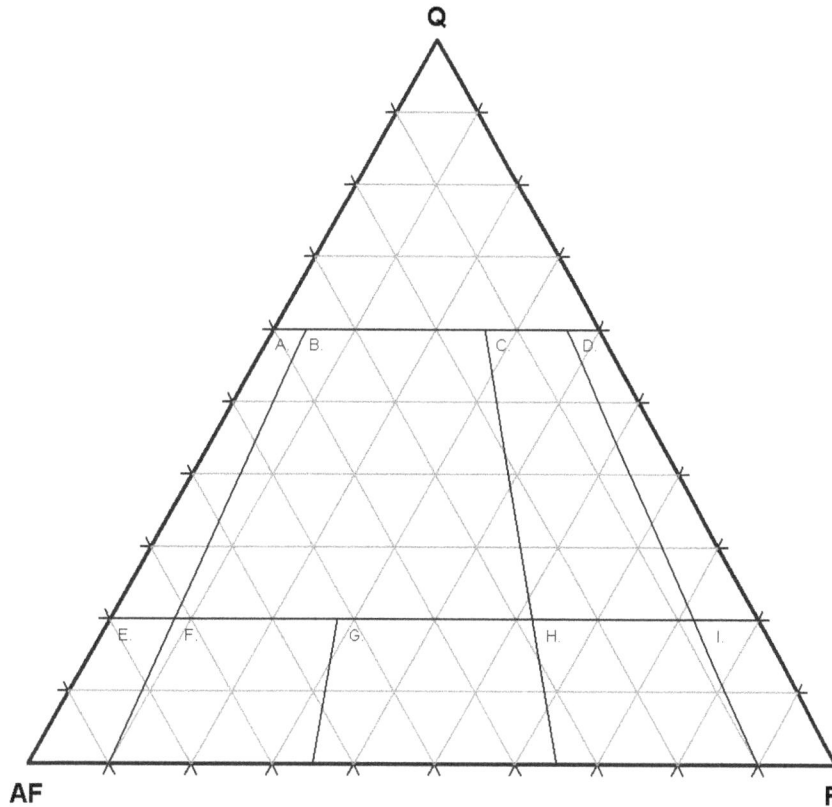

A. Alkali Feldspar Charnockite
B. Charnockite
C. Opdalite or Charno-enderbite
D. Enderbite
E. Hypersthene Alkali Feldspar Syenite
F. Hypersthene Syenite
G. Mangerite
H. Jotunite
I. Norite, Anorthosite (M<10%)

Definition: Contains >5% hypersthene

Melilite Plutonic Rocks

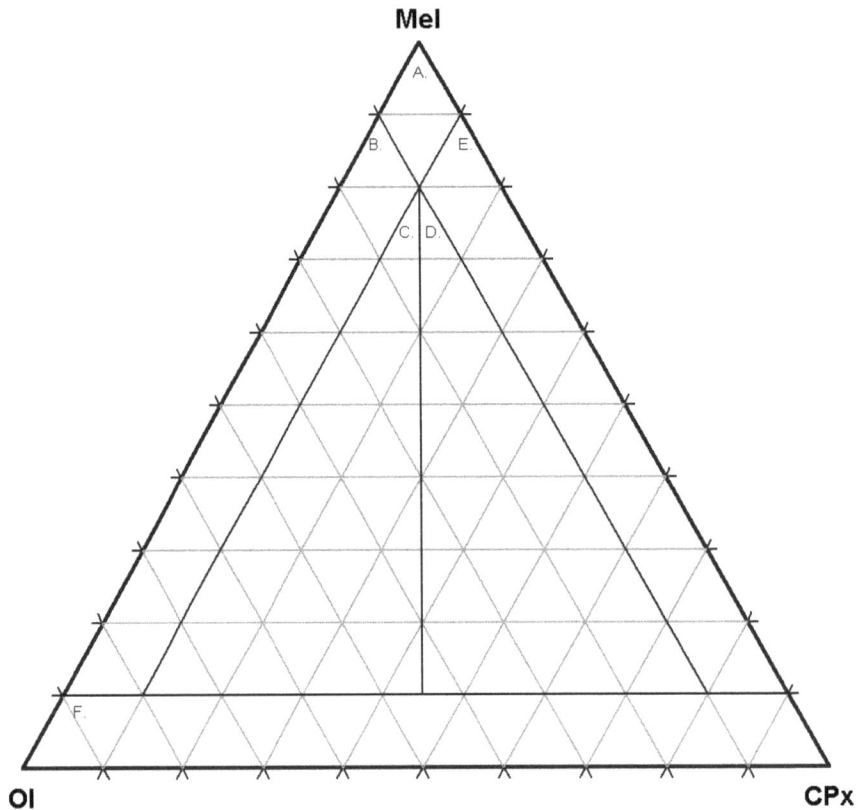

A. Melilitolite
B. Olivene Melilitolite or Kugdite
C. Pyroxene Olivene Melilitolite
D. Olvene Pyrexene Melitolite (Olivene Uncompahgrite)
E. Pyroxene Melilitolite (Uncompahgrite)
F. Melilite Bearing Peridotite (Olivene >40%)
G. Melilite Bearing Pyroxenite (Olivene <40%)

Definition: Contains >10% melilite and <10% feldspathoids

Melilite Volcanic Rocks

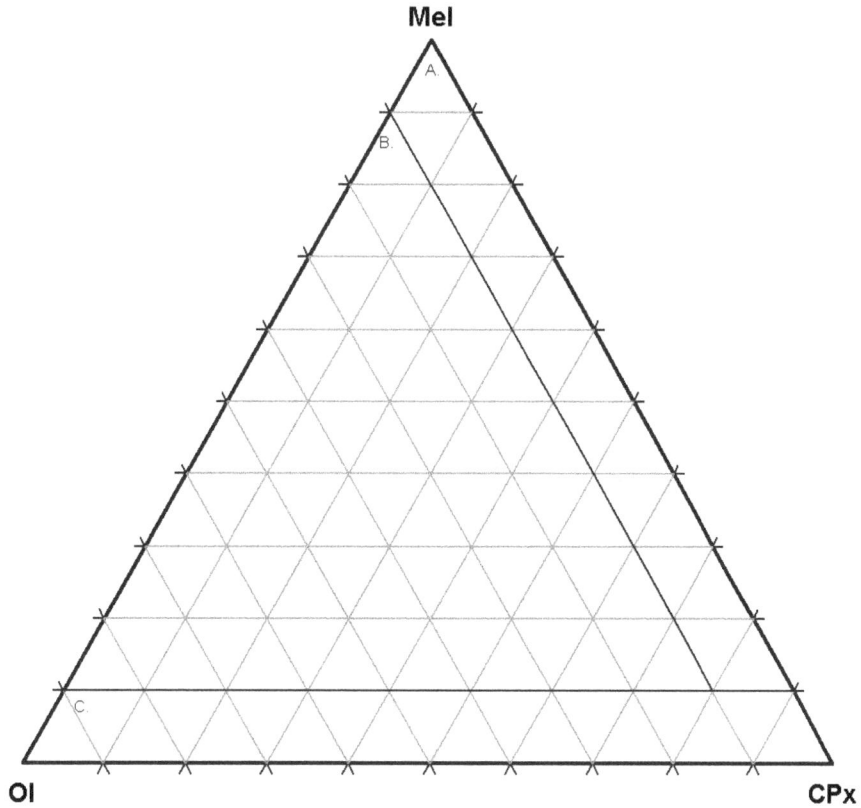

A. Melilite
B. Olivene Melilite
C. Melilite Bearing Pyroxenite (Olivene<40%)
D. Melilite Bearing Peridotite (Olivene >40%)

Definition: Contains >10% melilite and <10% feldspathoids

Plutonic Rocks QAPF

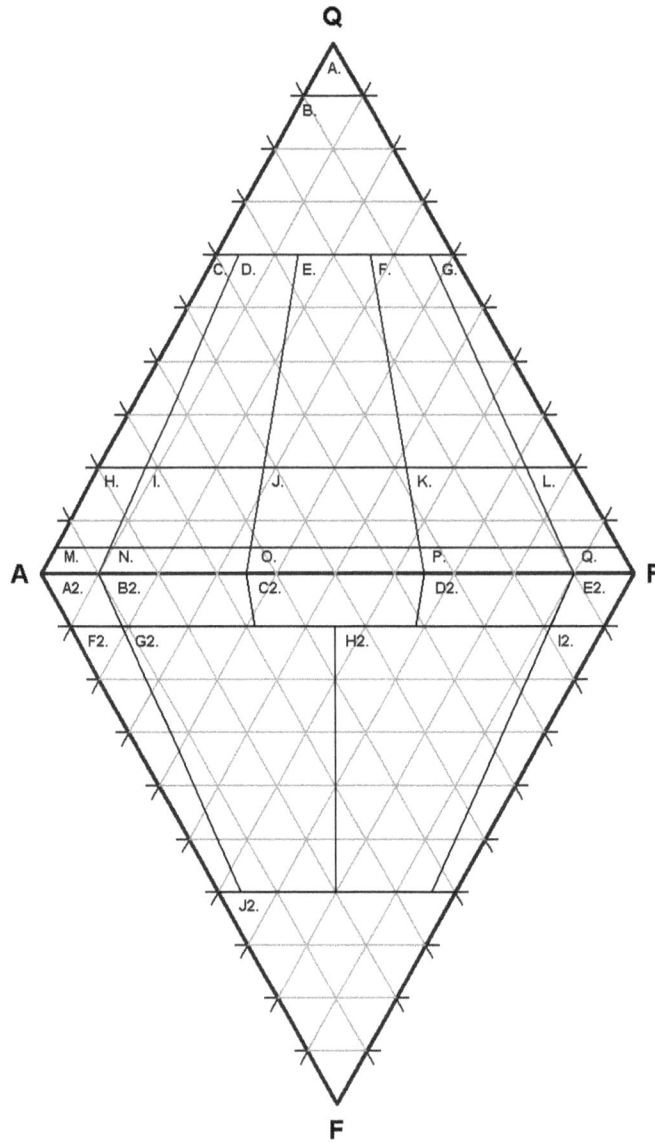

Plutonic Rocks QAPF Description

A. Quartzolite	Q. Diorite (An<50%)
B. Quartz Rich Granitoids	Q. Gabbro (An>50%)
C. Alkali Feldspar Granite	Q. Anorthosite (M<10%)
D. Syenogranite (Granite)	A2. Foid Bearing Alkali Feldspar Syenite
E. Monzogranite (Granite)	B2. Foid Bearing Syenite
F. Granodiorite	C2. Foid Bearing Monzonite
G. Tonalite	D2. Foid Bearing Monzodiorite (An<50%)
H. Quartz Alkali Feldspar Syenite	D2. Foid Bearing Monzogabbro (An>50%)
I. Quartz Syenite	E2. Foid Bearing Diorite (An<50%)
J. Quartz Monzonite	E2. Foid Bearing Gabbro (An>50%)
K. Quartz Monzodiorite (An<50%)	E2. Foid Bearing Anorthosite (M<10%)
K. Quartz Monzogabbro (An>50%)	F2. Foid Syenite
L. Quartz Diorite (An<50%)	G2. Foid Monzosyenite
L. Quartz Gabbro (An>50%)	H2. Foid Monzodiorite (An<50%)
L. Quartz Anorthosite (M<10%)	H2. Foid Monzogabbro (An>50%)
M. Alkali Feldspar Syenite	I2. Foid Diorite (An<50%)
N. Syenite	I2. Foid Gabbro (An>50%)
O. Monzonite	I2. Foid Anorthosite (M<10%)
P. Monzodiorite (An<50%)	J2. Foidolite
P. Monzogabbro (An>50%)	

Definition: Contains <90% mafic minerals

Gabbroic Rocks Plagioclase/Pyroxene/Olivene

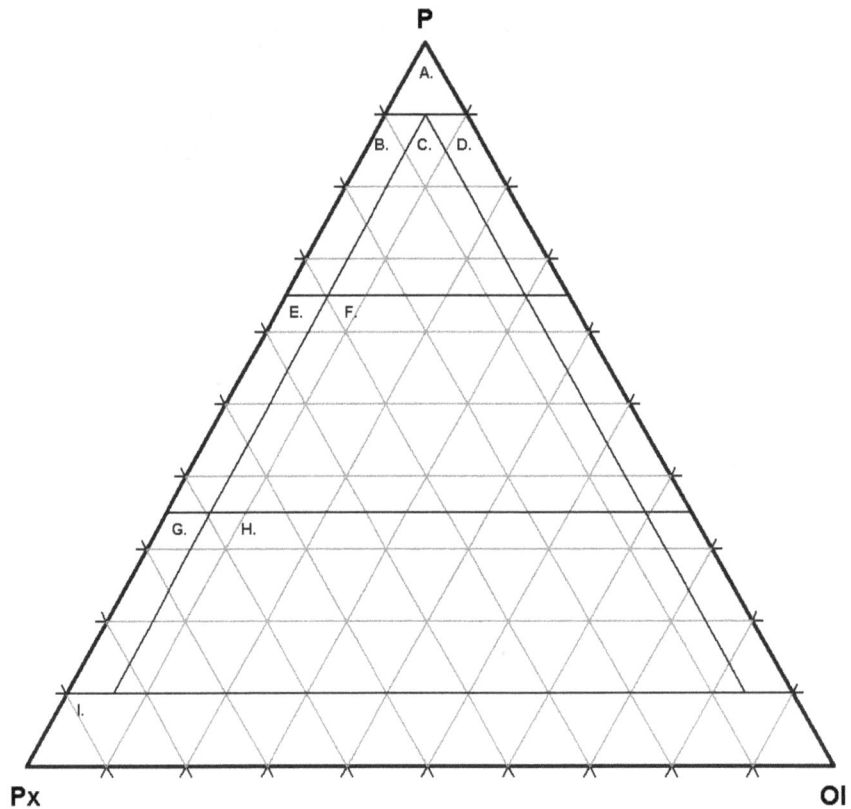

A. Anorthosite
B. Gabbro
C. Olivene Gabbro
D. Troctolite
E. Gabbronorite
F. Olivene Gabbronorite
G. Norite
H. Olivene Norite
I. Plagioclase Bearing Ultramafics

Definition: Falls into the gabbroic field of the QAPF diagram

Gabbroic Rocks Plagioclase/Orthopyroxene/Clinopyroxene

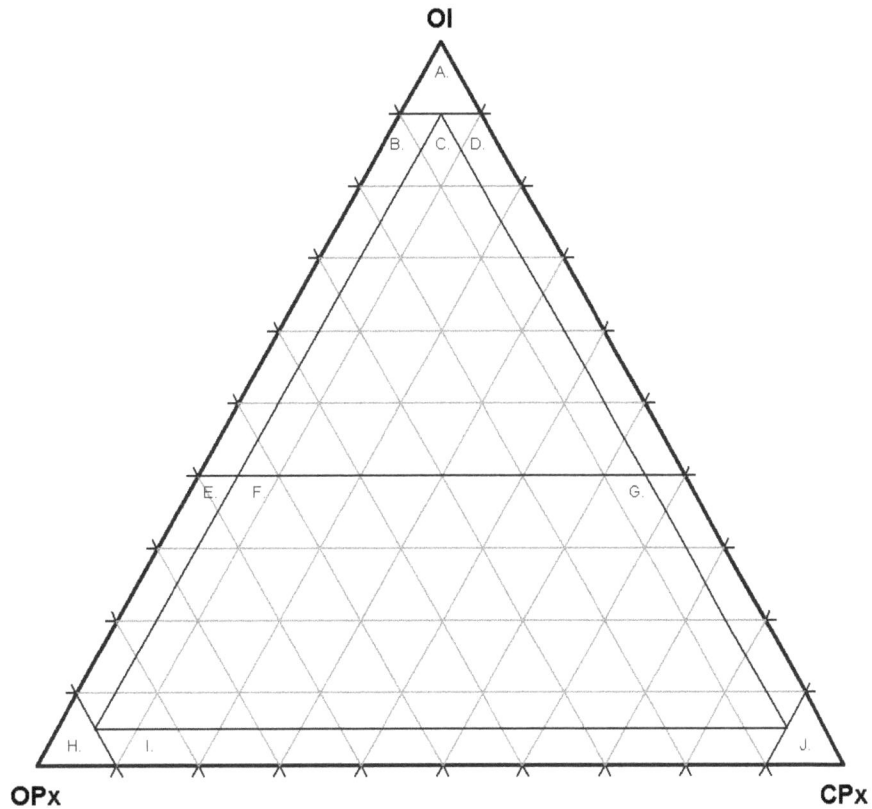

A. Anorthosite
B. Norite
C. Clinopyroxene Norite (Gabbronorite)
D. Orthopyroxene Gabbro (Gabbronorite)
E. Gabbro
F. Plagioclase Bearing Pyroxenite

Definition: Falls into the gabbroic field of the QAPF diagram

Gabbroic Rocks Plagioclase/Pyroxene/Hornblende

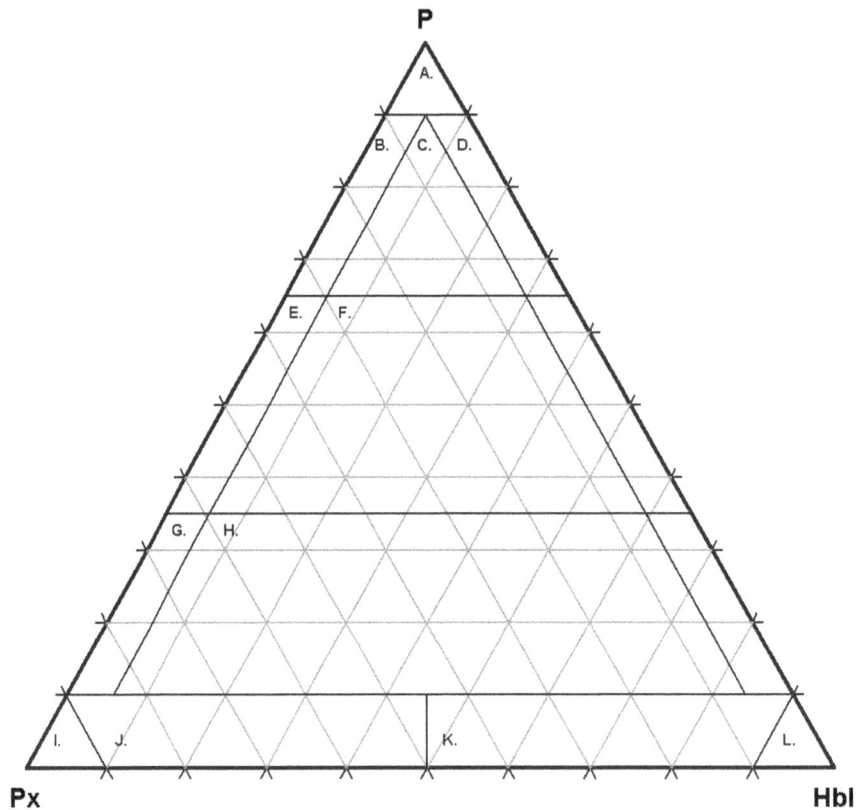

A. Anorthosite
B. Gabbro
C. Pyroxene Hornblende Gabbro
D. Hornblende Gabbro
E. Gabbronorite
F. Hornblende Gabbronorite
G. Norite
H. Pyroxene Hornblende Norite
I. Plagioclase Bearing Pyroxenite
J. Plagioclase Bearing Hornblende Pyroxenite
K. Plagioclase Bearing Pyroxene Hornblendite
L. Plagioclase Bearing Hornblendite

Definition: Falls into the gabbroic field of the QAPF diagram

Ultramafic Rocks Olivene/Orthopyroxene/Clinopyroxene

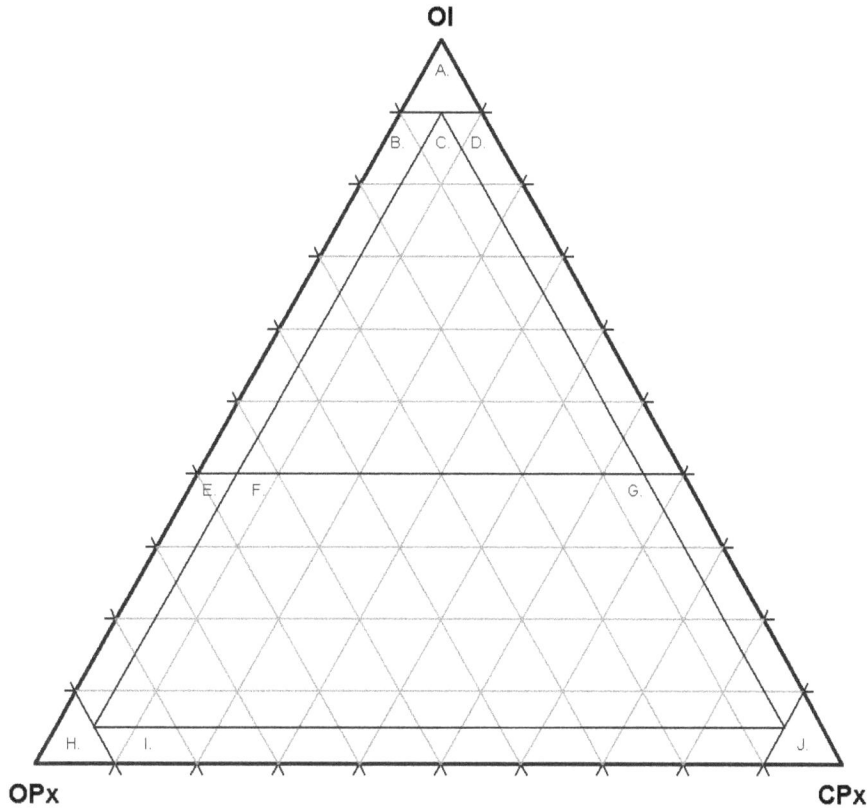

A. Dunite
B. Harzburgite
C. Lherzolite
D. Wehrlite
E. Olivene Orthopyroxenite
F. Olivene Websterite
G. Olivene Clinopyroxenite
H. Orthopyroxenite
I. Websterite
J. Clinopyroxenite

Definition: Contains >90% mafic minerals

Ultramafic Rocks Olivene/Pyroxene/Hornblende

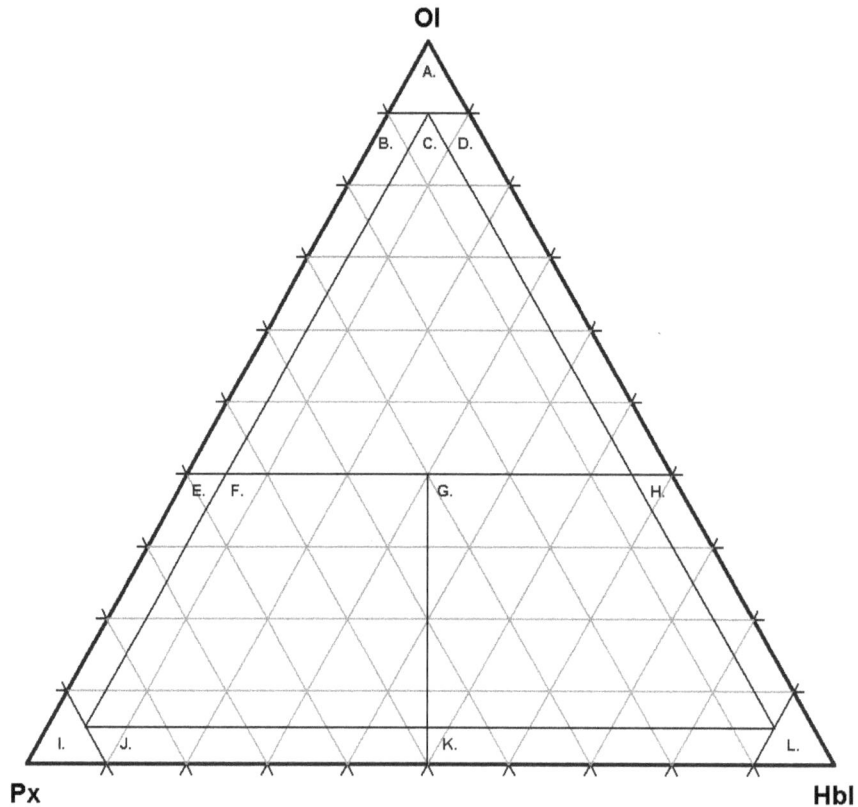

A. Dunite
B. Pyroxene Peridotite
C. Pyroxene Hornblende Peridotite
D. Hornblende Peridotite
E. Olivene Pyroxenite
F. Olivene Hornblende Pyroxenite
G. Olivene Pyroxene Hornblendite
H. Olivene Hornblendite
I. Pyroxenite
J. Pyroxene Hornblendite
K. Hornblende Pyroxenite
L. Hornblendite

Definition: Contains >90% mafic minerals

Ultramafic Rocks

Further Subdivisions	trachybasalt	basaltic trachyandesite	trachyandesite
Na2O - 2.0 >= K2O	hawaiite	mugearite	benmoreite
Na2O - 2.0 <= K2O	potassic trachybasalt	shoshonite	latite

Volcanic Rocks QAPF

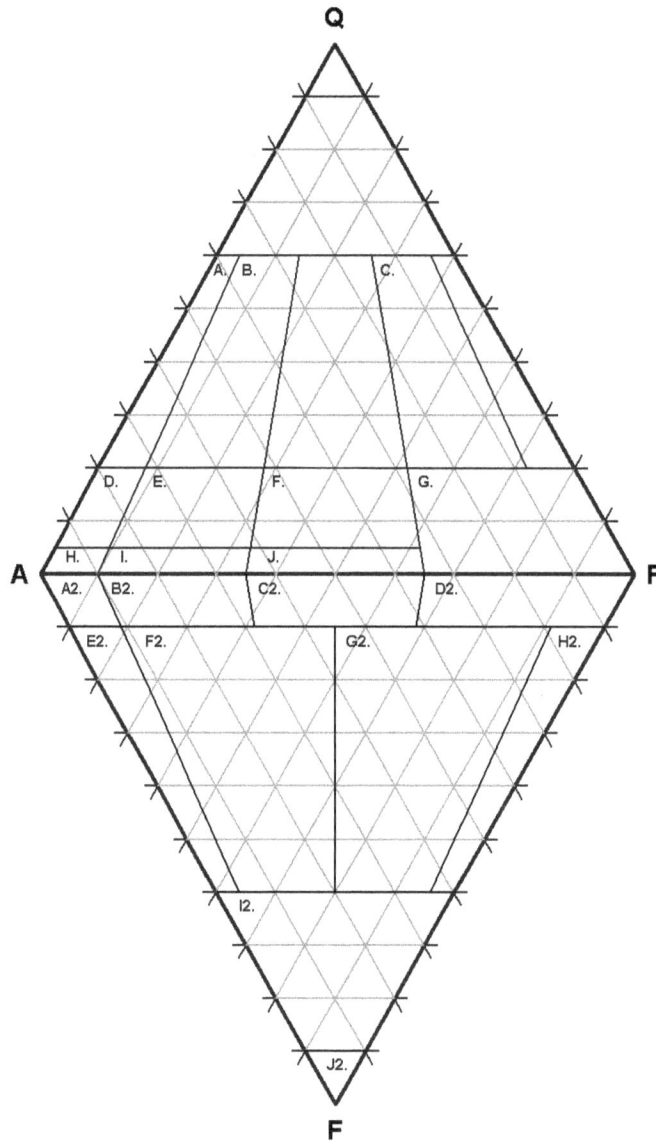

Volcanic Rocks QAPF Description

A. Alkali Feldspar Rhyolite	C2. Foid Bearing Latite
B. Rhyolite	D2. Basalt (M=35-90 or SiO2<52%)
C. Dacite	D2. Andesite (M<35% or SiO2>52%)
D. Quartz Alkali Feldspar Trachyte	E2. Phonolite
E. Quartz Trachyte	F2. Tephritic Phonolite
F. Quartz Latite	G2. Phonolitic Basanite (Olivene>10%)
G. Basalt (M=35-90 or SiO2<52%)	G2. Phonolitic Tephrite (Olivene<10%)
G. Andesite (M<35% or SiO2>52%)	H2. Basanite (Olivene>10%)
H. Alkali Feldspar Trachyte	H2. Tephrite (Olivene<10%)
I. Trachyte	I2. Phonolitic Foidite
J. Latite	J2. Tephritic Foidite
A2. Foid Bearing Alkali Feldspar Trachyte	K2. Foidite
B2. Foid Bearing Trachyte	

Definition: Contains <90% mafic minerals

Volcanic Rocks Total Alkali/Silica

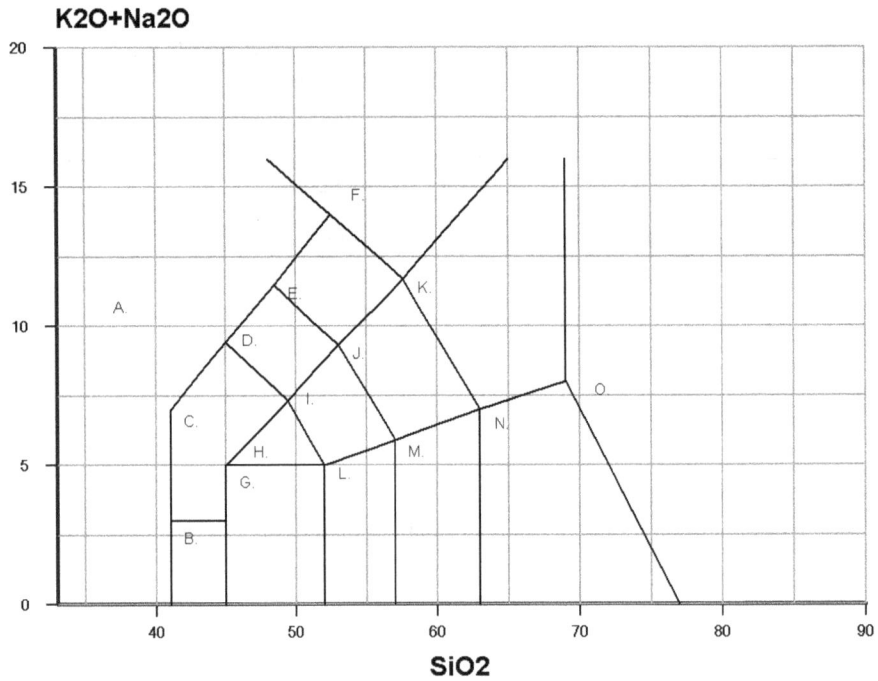

A. Foidite	I. Basaltic Trachyandesite
B. Picro Basalt	J. Trachyandesite
C. Basanite (Olivene >10%)	K. Trachyte (Quartz<20%)
C. Tephrite (Olivene<10%)	K. Trachydacite (Quartz>20%)
D. Phonotephrite	L. Basaltic Andesite
E. Tephrophonolite	M. Andesite
F. Phonolite	N. Dacite
G. Basalt	O. Rhyolite
H. Trachybasalt	

Lamprophyres

Felsic Mineral	Felsic Mineral	Predominant Mafic Minerals			
feldspar	foid	biotite diopside augite (olivene)	hornblende diopside augite (olivene)	amphibole titanaugite olivene biotite	matrix biotite (titanaugite) (olivene)
Color Index		>35%	>35%	>40%	>70%
or>pl pl>or absent pl>or absent absent	absent absent glass or foid fsp>foid foid foid	minette kersantite	vogesite spessartite	sannaite camptonite monchiquite	polzenite alnoite
Lamprophyre Type		Calc-Alkaline	Calc-Alkaline	Alkaline	Melilitic

Pyroclastics

Clast Size in mm	Pyroclast	Pyroclastc Deposit	
		Mainly Unconsolidated Tephra	Mainly Consolidated Pyroclastic Rock
>64mm	bomb, block	agglomerate bed of blocks or bomb, block tephra	agglomerate pyroclastic breccia
<64mm and >2mm	lapillus	layer, bed of lapilli or lapilli tephra	lapilli tuff
<2mm and >1/16mm	coarse ash grain	coarse ash	coarse (ash) tuff
<1/16mm	fine ash grain (dust grain)	fine ash (dust)	fine (ash) tuff (dust tuff)

Metamorphic Rocks

Dynamometamorphic Rocks

Nature of Matrix	Proportion of Matrix			
	0-10%	10-50%	50-90%	90-100%
crushed foliated	tectonic breccia	protomylonite	mylonite	ultramylonite
crushed massive	or conglomerate	protocataclasite	cataclasite	ultracataclasite
recystallized minor	hartschelfer			
glassy	blastomylonite			
recrystallized major	pseudotachylite			

Foliated Rocks

Structure		Rockname
Schistose		Schist
Slaty		Phyllite, Slate
Foliated/Banded		Gneiss
Mixed		Migmatite
Crushed and/or recrystallized		Mylonite

Crystalline Metamorphic Rocks

Mineralogy		Rockname
Hornblende, plagioclase, silliminite, andalusite, cummingtonite, biotite, staurolite, diopside		Amphibolite
Calcite, dolomite, brucite		Marble
Diopside, wollastonite, grossular, zoisite, vesuvianite		Skarn
Serpentine, olivene, pyroxene		Serpentinite
Quartz, plagioclase, garnet, silliminite, biotite (granulitic)		Granulite
Plagioclase, orthoclase, quartz hypersthene, biotite (charnockitic)		Charnockite
Clinopyroxene, garnet		Eclogite

Sedimentary Rocks

Sandstones Quartz/Clay/Feldspar

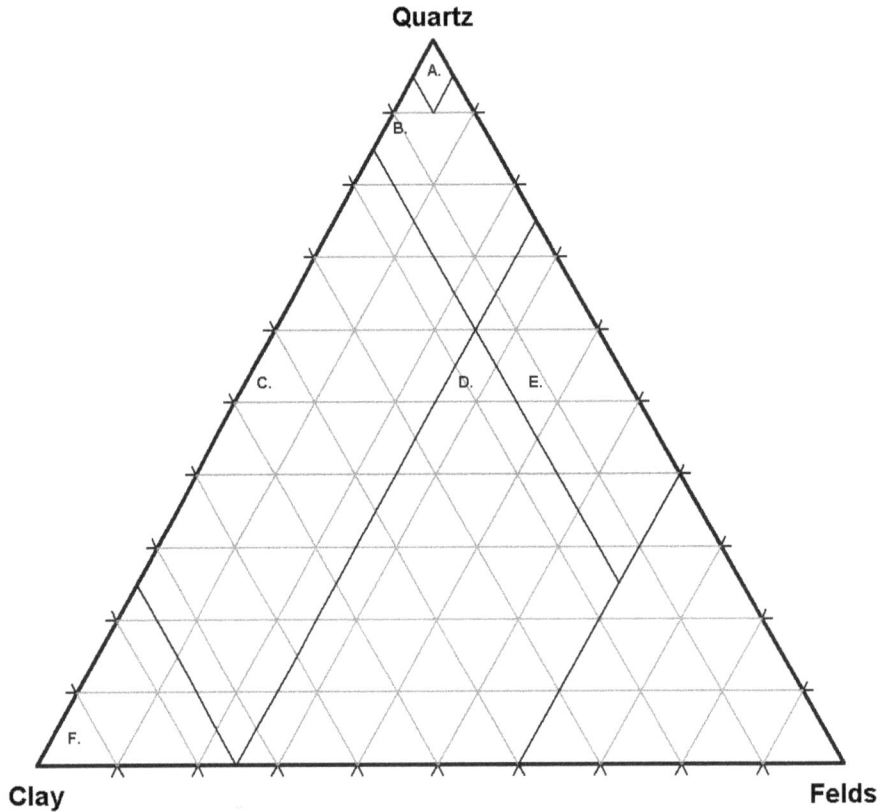

A. Quartz Arenite (Orthoquartzite)
B. Sub Arkose (Protoquartzite)
C. Quartz Wacke
D. Greywacke
E. Arkose
F. Mudrocks (Clay>75%)
C., D. Wackes (Clay >15% and <75%)
A., B., E. Arenites (Clay <15%)

Sandstones Quartz/Feldspar/Rock Fragments

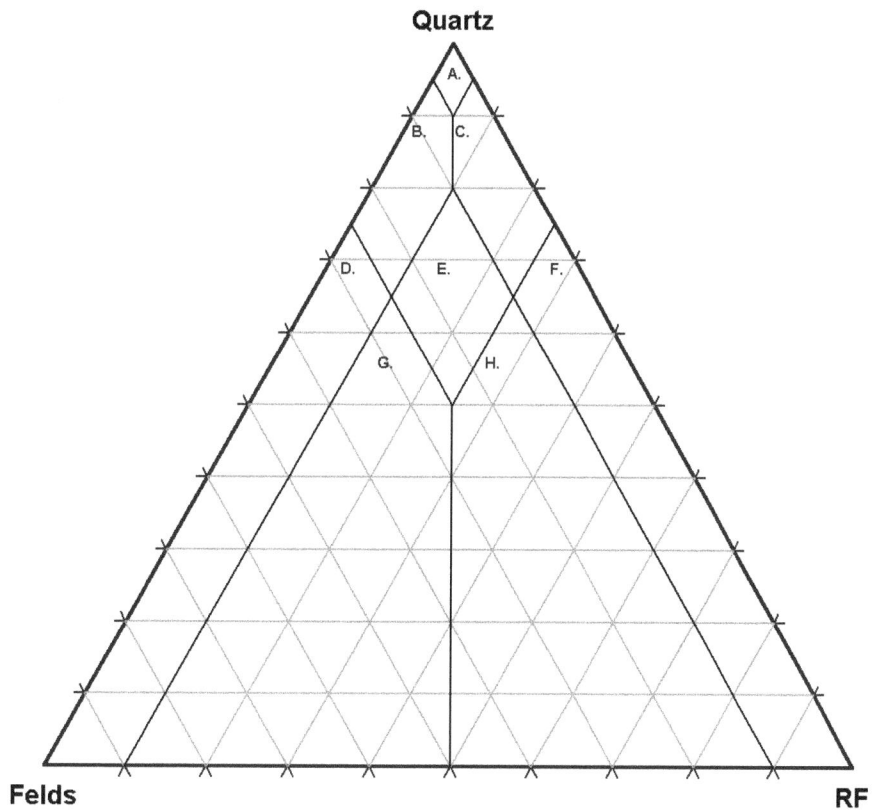

A. Quartz Arenite
B. Subarkose
C. Sublitharenite
D. Arkose
E. Lithic Subarkose
F. Litharenite
G. Lithic Arkose
H. Feldspathic Litharenite

Sandstones-Arenites

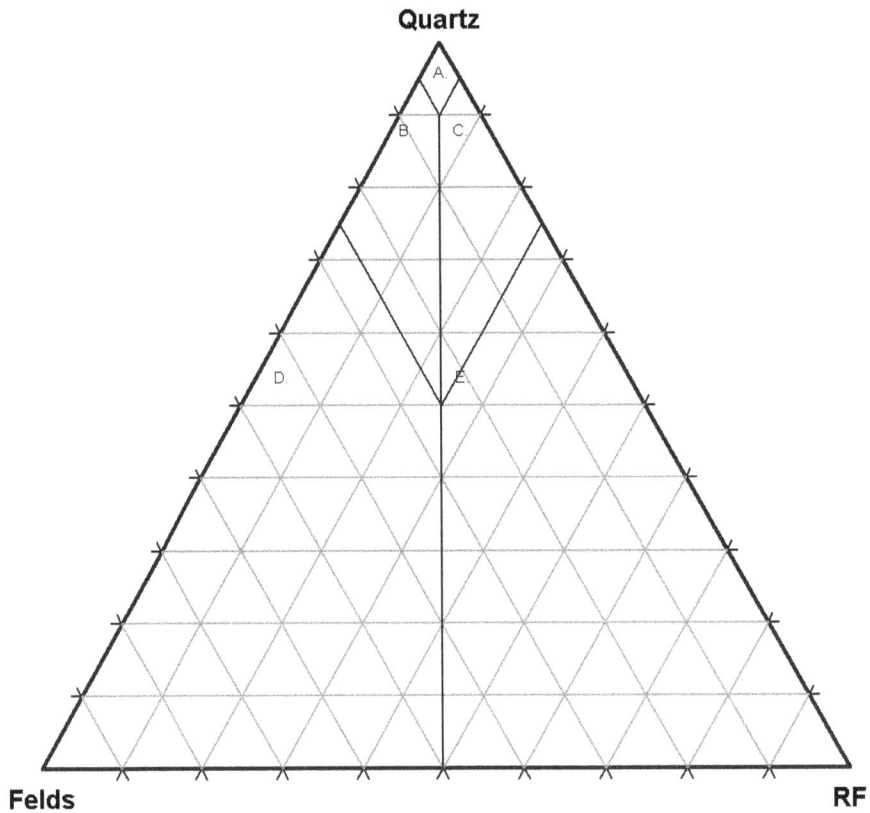

A. Quartz Arenite
B. Subarkose
C. Sublitharenite
D. Arkosic Arenite or Arkose (Feldspar >75%)
E. Arkosic Arenite or Lithic Arkose (Feldspar<75%)
F. Litharenite

Definition: Less than 15% matrix

Sandstones-Wackes

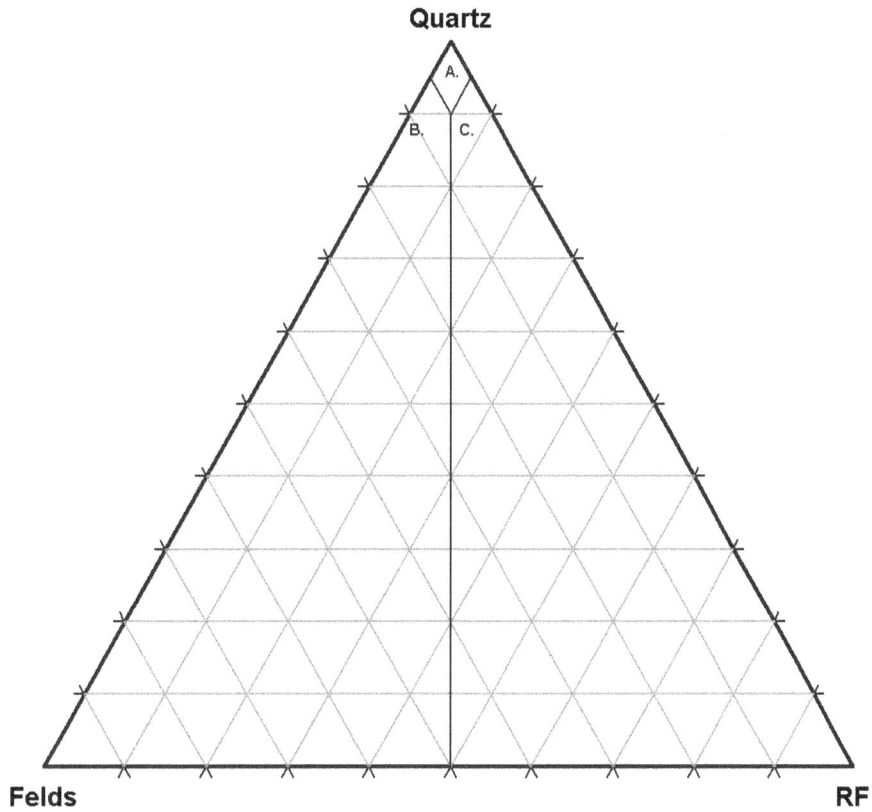

A. Quartz Wacke
B. Feldspathic Wacke
C. Lithic Wacke

Definition: 15% to 75% matrix

Conglomerates

Epiclastic	extraformational	orthoconglomerates matrix <15%	metastable <10% orthoquartzitic oligomict conglomerate
			metastable >10% petromict conglomerate
		paraconglomerates or diamictites matrix >15%	laminated matrix: lam. conglomeratic mudstone
			non laminated matrix: tillite (glacial) tilloid (nonglacial)
	intraformational	intraformational conglomerates and breccias	
Pyroclastic	volcanic breccias and agglomerates		
Cataclastic	landslide and slump breccias fault and fold breccias (tectonic moraines) collapse and solution breccias		
Meteoric	impact breccias		

Limestones

1. Original Components Not Organically Bound					
of the allochems less than 10% >2mm				of the allochems more than 10% >2mm	
contains carbonate mud particles <0.03mm diameter			mud absent	matrix supported	grain supported
mud supported		grain supported		matrix supported	grain supported
less than 10% grains	more than 10% grains	grain supported		matrix supported	grain supported
mudstone	wackestone	packstone	grainstone	floatstone	rudstone
2. Original Components Organically Bound					
boundstone					
organisms acting as baffles		organisms encrusting and binding		organisms binding a rigid framework	
bafflestone		bindstone		framestone	

Principle Allochems in Limestone	Cemented by Grains	With a Micritic Matrix
>25% bioclasts	biosparite	biomicrite
>25% ooids	oosparite	oomicrite
>25% peloids	pelsparite	pelmicrite
>25% intraclasts	intrasparite	intramicrite

Mudstones

% clay size constituents		0-32%	32-65%	66-100%
non-indurated	beds >1cm	bedded silt	bedded mud	bedded claymud
	laminae < 1cm	laminated silt	laminated mud	laminated claymud / claystone
indurated	mud >1cm	bedded siltstone	mudstone	claystone
	laminae <1cm	laminated siltstone	mudshale	clayshale
meta-morphosed	increasing metamorphism	quartz argillite	argillite	
		quartz slate	slate	
		phyllite and/or mica schist		

Phosphorites

Less than 10% grains < 10 microns	Phosphate mineralogy uncertain	Microsphorite
	phosphate mineral is apatite	Microsphatite
more than 10% grains > 10 microns	10 microns < grains <63 microns	Phosphalutite
	63 microns < grains < 2mm	Phospharenite
	grains > 2mm	Phospharudite

Volcaniclastics

Clast Size in mm	Pyroclastic	Volcaniclastics	Sedimentary
>64 mm	agglomerate pyroclastic breccia	tuffaceous conglomerate tuffaceous breccia	conglomerate breccia
>2mm and <64mm	lapilli tuff		
>1/16mm and <2mm	(ash) tuff coarse	tuffaceous sandstone	Sandstone
>126mm and <1/16mm	(ash) tuff fine	tuffaceous siltstone	siltstone
< 1/256mm	(ash) tuff	tuffaceous mudstone, shale	mudstone, shale
Pyroclastic Material	100% to 75%	75% to 25%	25% to 0%

Terminology

These are the basic petrological terms used for the rock descriptions.

Group

Igneous – Formed by crystallization out of magma derived from the interior of the Earth
Metamorphic – Formed by the recrystallization of pre-existing rocks due to increases in temperature and/or pressure
Sedimentary – 1.Formed by the mechanical accumulation of grains from pre-existing rocks. 2.Formed by the chemical precipitation of minerals 3.Formed by the burial and maturation of organic matter

Structure

Massive - No structure
Foliated - Planar alignment of platy minerals
Schistose - Discontinuous planar alignment of minerals
Flow - Changing alignment of minerals in a flow pattern
Banded - With bands of different composition or color
Bedded/Laminated - Divided into separate layers

Texture

Crystalline - All crystalline textures, with a grain size > 0.2mm (e.g. Granite, Gneiss)
Microcrystalline - Crystalline textures with a grain size < 0.2mm and > 0.01mm (e.g. Hornfels)
Cryptocrystalline - Crystalline textures with a grain size < 0.01mm (e.g. Agate)
Amorphous - Non crystalline, glassy (e.g. Obsidian)
Porphyritic - Large crystals (phenocrysts) in a fine-grained ground mass (e.g. Basalt)
Recrystallized - Textures produced by partial recrystallization (e.g. Meta-Basalt)
Inequigranular - Fine to macro - sized grains (e.g. Kimberlite)
Fragmental - Composed of mineral and/or rock fragments (e.g. Pyroclastics)
Biogenic- Textures produced by organisms (e.g. Limestone - Boundstone)
Organic-Textures produced by organic material (e.g. coal)
Clastic - Textures produced by mechanically accumulated grains cemented together (e.g. Sandstone)
Chemical - Textures produced by chemical precipitation (e.g. Anhydrite)

Occurrence

Plutonic - A general term for any large scale intrusive rocks
Volcanic - Extrusive and associated intrusive rocks
Regional Metamorphic - Metamorphic rocks occurring over large areas
Contact Metamorphic - Adjacent to intrusions
Fault/Shear Zone - Planar zones of brittle and/or ductile deformation
Basin - Sedimentary rocks in a sedimentary basin.

Rock Descriptions

Rockname: Achondrites

Group: Meteorites

Family: Stony meteorites

Texture: Crystalline

Structure: Massive

Composition: Pyroxenes, olivene, plagioclase

Description: A general term for stony meteorites without chondrules

Rockname: Actinolite Schist

Group: Metamorphic

Family:

Texture: Crystalline, microcrystalline

Structure: Schistose

Composition: Actinolite

Description: Formed by low grade regional metamorphism of mainly pelitic sediments

Rockname: Agate

Group:

Family: Agates

Texture: Cryptocrystalline

Structure: Banded

Composition: Chalcedony

Description: Formed from silica gels in cavities of volcanic rocks. Diffusion of metallic impurities into layers of different porosity produce the colored bands

Rockname: Agglomerate

Group: Igneous

Family: Pyroclastic

Texture: Fragmental, unconsolidated, grain size > 64mm

Structure: Massive, bedded

Composition: Rock fragments, crystal fragments, glass fragments

Description: Ejecta accumulated by volcanic activity

Rockname: Alaskite

Group: Igneous

Family: Granites

Texture: Crystalline

Structure: Massive

Composition: Quartz, orthoclase (K feldspar), minor biotite, hornblende (amphibole)

Description: A variety of Alkali Feldspar Granite with minor mafic minerals.

Rockname: Albite Epidote Hornfels Facies

Group: Metamorphic

Family:

Texture: Microcrystalline

Structure: Massive

Composition: Plagioclase, epidote, biotite

Description: A metamorphic facies containing rocks that have undergone low to medium grade contact metamorphism

Rockname: Algal Boundstone

Group: Sedimentary

Family: Limestones

Texture: Biogenic

Structure: Massive, bedded, laminated

Composition: Algae, micrite, sparite, calcite

Description: A limestone bound together by algae

Rockname: Alkali Basalt

Group: Igneous

Family: Basalts

Texture: Porphyritic

Structure: Massive, flow structure

Composition: Augite (clinopyroxene), nepheline (foids), alkali amphibole, olivene, plagioclase, glass

Description: A Basaltic rock of alkali composition that occurs as lava flows and intrusive rocks

Rockname: Alkali Feldspar Charnockite

Group: Igneous, metamorphic

Family: Charnockites

Texture: Crystalline

Structure: Massive

Composition: Orthoclase (K feldspar), hypersthene, quartz, biotite, hornblende (amphibole)

Description: A granitoid containing hypersthene

Rockname: Alkali Feldspar Granite

Group: Igneous

Family: Granites

Texture: Crystalline

Structure: Massive

Composition: Quartz, orthoclase (K feldspar), biotite, hornblende (amphibole)

Description: A variety of granite

Rockname: Alkali Feldspar Rhyolite

Group: Igneous

Family: Rhyolites

Texture: Porphyritic

Structure: Massive, flow structure

Composition: Sanidine (K feldspar), quartz, biotite, minor mafic minerals

Description: Lava flows and intrusives of felsic composition

Rockname: Alkali Feldspar Syenite

Group: Igneous

Family: Syenites

Texture: Crystalline

Structure: Massive

Composition: Orthoclase (K feldspar), biotite, hornblende (amphibole)

Description: Dominant felsic mineral is K-feldspar. Lacks quartz

Rockname: Alkali Feldspar Trachyte

Group: Igneous

Family: Trachytes

Texture: Porphyritic

Structure: Massive, flow structure

Composition: Sanidine (K feldspar), biotite, hornblende (amphibole), pyroxenes

Description: A felsic volcanic rock with K feldspar the major felsic mineral. Occurs as lava flows and intrusions

Rockname: Alnoite

Group: Igneous

Family: Lamprophyres

Texture: Porphyritic

Structure: Massive

Composition: Melilite, biotite, augite, olivene, calcite

Description: A mafic to ultramafic minor intrusive rock

Rockname: Alvikite

Group: Igneous

Family: Carbonatites

Texture: Crystalline

Structure: Massive

Composition: Calcite

Description: Consisting primarily of calcite

Rockname: Amphibolite

Group: Metamorphic

Family:

Texture: Crystalline

Structure: Massive, foliated

Composition: Plagioclase, hornblende, silliminite, andalusite, cordierite, cummingtonite, staurolite, biotite, diopside (clinopyroxene)

Description: A medium grade metamorphic rock composed of hornblende and plagioclase

Rockname: Amphibolite Facies

Group: Metamorphic

Family:

Texture: Crystalline

Structure: Massive, foliated, schistose

Composition: Plagioclase, hornblende, silliminite, andalusite, cordierite, cummingtonite, staurolite, biotite, diopside (clinopyroxene)

Description: A metamorphic facies containing rocks that have undergone medium grade regional metamorphism

Rockname: Analcimite

Group: Igneous

Family: Foidites

Texture: Porphyritic

Structure: Massive

Composition: Analcime (foids), pyroxenes, amphibole, biotite, olivene

Description: Alkali volcanic rock with analcime the dominant felsic mineral

Rockname: Andesite

Group: Igneous

Family: Andesites

Texture: Porphyritic

Structure: Massive, flow structure

Composition: Plagioclase, pyroxenes, hornblende (amphibole), biotite, mafics < 35%

Description: Lava flows and intrusive rocks of intermediate composition associated with subducting plate margins (island arcs)

Rockname: Angrites

Group: Meteorites

Family: Stony meteorites

Texture: Crystalline

Structure: Massive

Composition: Augite

Description: Found after meteorite showers

Rockname: Anhydrite

Group: Sedimentary

Family: Evaporites

Texture: Chemical, nodular

Structure: Massive, bedded, laminated

Composition: Anhydrite, gypsum, matrix

Description: Formed by precipitation during evaporation of brines

Rockname: Anorthosite

Group: Igneous

Family: Anorthosites

Texture: Crystalline

Structure: Massive

Composition: Plagioclase

Description: A plutonic rock consisting of plagioclase

Rockname: Anthracite

Group: Sedimentary

Family: Coals

Texture: Organic

Structure: Massive, bedded, laminated

Composition: Vitrinite, inertinite, liptonite, mineral matter

Description: A high grade (high carbon content) coal

Rockname: Arenite

Group: Sedimentary

Family: Sandstones

Texture: Clastic, grains 0.0625 to 2mm

Structure: Massive, bedded, laminated

Composition: Quartz, orthoclase (K feldspar) plagioclase, rock fragments, clay < 15%

Description: A sandstone with minor clay matrix

Rockname: Argillite

Group: Metamorphic

Family:

Texture: Recystallized

Structure: Massive, bedded, laminated

Composition: Clay

Description: A hard, compact rock formed by low grade metamorphism of argillaceous sediments.

Rockname: Arkose

Group: Sedimentary

Family: Sandstones

Texture: Clastic, grains 0.0625 to 2mm

Structure: Massive, bedded, laminated

Composition: Quartz, orthoclase (K feldspar) plagioclase, clay < 15%

Description: A feldspathic sandstone with < 15% clay matrix

Rockname: Arkosic Arenite

Group: Sedimentary

Family: Sandstones
Texture: Clastic, grains 0.0625 to 2mm
Structure: Massive, bedded, laminated
Composition: Quartz, orthoclase (K feldspar) plagioclase, rock fragments, clay < 15%
Description: A feldspathic sandstone with < 15% clay matrix

Rockname: Ash
Group: Igneous
Family: Tephra
Texture: Fragmental, unconsolidated, grain size < 2mm
Structure: Massive, bedded, laminated
Composition: Rock fragments, crystal fragments, glass fragments
Description: Ejecta accumulated by volcanic activity

Rockname: Atexite
Group: Meteorites
Family: Iron meteorites
Texture: Crystalline
Structure: Massive
Composition: Fe-Ni alloys, troilite, graphite
Description: Found after meteorite showers

Rockname: Aubrites
Group: Meteorites
Family: Stony meteorites
Texture: Crystalline
Structure: Massive
Composition: Enstatite
Description: Found after meteorite showers

Rockname: Australites
Group: Tektites

Family:

Texture: Amorphous

Structure: Massive

Composition: Glass

Description: A term for tektites found in Australia

Rockname: Bafflestone

Group: Sedimentary

Family: Limestones

Texture: Biogenic

Structure: Massive, bedded, laminated

Composition: Fossils, algae, micrite, sparite, calcite

Description: A type of boundstone with organisms acting as baffles

Rockname: Banded Iron Formation

Group: Sedimentary

Family: Ironstones

Texture: Cryptocrystalline, microcrystalline

Structure: Banded

Composition: Hematite, magnetite, greenalite, grunerite, chert

Description: Deposited by chemical precipitation from sea water. Iron oxides are probably formed by bacterial activity

Rockname: Basalt

Group: Igneous

Family: Basalts

Texture: Porphyritic

Structure: Massive, flow structure

Composition: Plagioclase, augite (clinopyroxene), olivene, hornblende, orthopyroxene, glass

Description: A general term for mafic volcanic rocks with plagioclase as the major felsic mineral. Occurs as lava flows

Rockname: Basaltic Trachyandesite

Group: Igneous

Family: Basalts

Texture: Porphyritic

Structure: Massive, flow structure

Composition: Plagioclase, augite (clinopyroxene), sanidine (K feldspar), hornblende, glass

Description: A variety of Basalt of intermediate composition that falls between trachyandesite and trachybasalt.

Rockname: Basanite

Group: Igneous

Family: Basanites

Texture: Porphyritic

Structure: Massive, flow structure

Composition: Clinopyroxene, plagioclase, foids, olivene

Description: A volcanic rock of alkali composition occurring as lava flows and intrusions

Rockname: Bauxite

Group: Sedimentary

Family:

Texture: Chemical, pisolitic

Structure: Massive

Composition: Bauxite, Aluminium oxides

Description: Precipitated on and near the surface from ground waters due to weathering of granites

Rockname: Beforsite

Group: Igneous

Family: Carbonatites

Texture: Crystalline

Structure: Massive

Composition: Dolomite

Description: Consisting primarily of dolomite

Rockname: Benmoreite

Group: Igneous

Family: Andesites

Texture: Porphyritic

Structure: Massive, flow structure

Composition: Anorthoclase (K feldspar) or sodic sanidine, olivene, augite

Description: A sodic variety of Trachyandesite defined chemically with $Na_2O - 2.0 >= K_2O$

Rockname: Bindstone

Group: Sedimentary

Family: Limestones

Texture: Biogenic

Structure: Massive, bedded, laminated

Composition: Fossils, algae, micrite, sparite, calcite

Description: A type of boundstone with organisms binding the framework

Rockname: Biolithite

Group: Sedimentary

Family: Limestones

Texture: Biogenic

Structure: Massive, bedded

Composition: Fossils, micrite, sparite, calcite

Description: A limestone organically bound and cemented by sparite

Rockname: Bituminous Coal

Group: Sedimentary

Family: Coals

Texture: Organic

Structure: Massive, bedded, laminated

Composition: Vitrinite, inertinite, liptonite, mineral matter
Description: A medium grade coal

Rockname: Blastomylonite

Group: Metamorphic
Family:
Texture: Fragmental, recystallized
Structure: Massive, foliated
Composition: Quartz, plagioclase, orthoclase (K feldspar), garnet, biotite, chlorite, epidote, calcite, crushed matrix, rock fragments, mineral fragments
Description: Composed of crushed and fragmental material that has undergone major recrystallization

Rockname: Boghead Coal

Group: Sedimentary
Family: Coals
Texture: Organic
Structure: Massive, bedded, laminated
Composition: Algal and minor fungal material
Description: A saprolitic coal

Rockname: Boninite

Group: Igneous
Family:
Texture: Porphyritic
Structure: Massive, flow structure
Composition: Enstatite (orthopyroxene), augite (clinopyroxene) in a glassy groundmass of crystallites
Description: A high magnesium glassy ultramafic lava

Rockname: Boundstone

Group: Sedimentary

Family: Limestones

Texture: Biogenic

Structure: Massive, bedded, laminated

Composition: Fossils, algae, micrite, sparite, calcite

Description: A general term for limestone that is organically bound

Rockname: Breccia

Group: Sedimentary

Family: Conglomerates

Texture: Clastic, grains > 4mm

Structure: Massive, bedded

Composition: Quartz, rock fragments, matrix

Description: Composed of angular, broken rock fragments. Deposited in rivers, alluvial fans and on scree slopes

Rockname: Calcarenite

Group: Sedimentary

Family: Limestones

Texture: Clastic

Structure: Massive, bedded, laminated

Composition: Fossils, micrite, sparite, calcite, quartz, >10% grains

Description: An old term for limestones containing >10% grains

Rockname: Calcilutite

Group: Sedimentary

Family: Limestones

Texture: Clastic

Structure: Massive, bedded, laminated

Composition: Micrite, <10% grains

Description: An old term for carbonate mudstones

Rockname: Calcirudite

Group: Sedimentary

Family: Limestones

Texture: Clastic, grains >2mm

Structure: Massive, bedded

Composition: Fossils, clasts, micrite, sparite, calcite, quartz

Description: An old term for very coarse grained and conglomeratic limestones

Rockname: Calcite carbonatite

Group: Igneous

Family: Carbonatites

Texture: Crystalline, microcyrstalline

Structure: Massive

Composition: Calcite

Description: Carbonate rich (> 50%) igneous rocks

Rockname: Calcrete

Group: Sedimentary

Family: Limestones

Texture: Cryptocrystalline

Structure: Massive, beeded, laminated

Composition: Calcite

Description: Formed by precipitation of calcite on and near the surface from ground waters. Stalactites, stalagmites and solution pipes consist of calcrete.

Rockname: Camptonite

Group: Igneous

Family: Lamprophyres

Texture: Porphyritic

Structure: Massive

Composition: Barkevite, kaersutite (amphibole), augite, olivene, biotite, plagioclase, foids, plag > or, felds > foids

Description: A mafic to ultramafic minor intrusive rock

Rockname: Cancarixite

Group: Igneous

Family: Lamproites

Texture: Inequigranular, porphyritic

Structure: Massive

Composition: Phlogopite, sanidine (K feldspar), diopside

Description: A type of lamproite

Rockname: Cannel Coal

Group: Sedimentary

Family: Coals

Texture: Organic

Structure: Massive, bedded, laminated

Composition: Vegetable matter, spores, algae, fungae

Description: A saprolitic coal

Rockname: Carbonaceous Chondrite

Group: Meteorites

Family: Stony meteorites

Texture: Crystalline, chondrules

Structure: Massive

Composition: Serpentine/olivene, chondrules (mineral granules), ferrite, organic matter

Description: A general term for carbonaceous chondrites

Rockname: Carbonatite

Group: Igneous

Family: Carbonatites

Texture: Crystalline, microcyrstalline

Structure: Massive

Composition: Dolomite

Description: A general term for igneous rocks containing > 50% carbonates

Rockname: Cataclasite

Group: Metamorphic

Family:

Texture: Fragmental, crushed matrix

Structure: Massive

Composition: Crushed matrix, rock fragments, mineral fragments

Description: Composed of fragmental material in a crushed matrix formed by medium to high grade pressure metamorphism

Rockname: Cedricite

Group: Igneous

Family: Lamproites

Texture: Inequigranular, porphyritic

Structure: Massive

Composition: Diopside, leucite

Description: A type of lamproite

Rockname: Chalk

Group: Sedimentary

Family: Limestones

Texture: Organic

Structure: Massive, bedded, laminated

Composition: Fossils

Description: A soft, white calcareous rock formed by the accumulation of minute calcareous fossils in ocean basins

Rockname: Charno-Enderbite

Group: Igneous, metamorphic

Family: Charnockites

Texture: Crystalline

Structure: Massive

Composition: Plagioclase, orthoclase (K feldspar), quartz, hypersthene, biotite, hornblende (amphibole)

Description: A granitoid containing hypersthene. Another name for Opdalite

Rockname: Charnockite

Group: Igneous, metamorphic

Family: Charnockites

Texture: Crystalline

Structure: Massive

Composition: Orthoclase (K feldspar), hypersthene, plagioclase, quartz, biotite, hornblende (amphibole)

Description: A granite containing hypersthene

Rockname: Chassignites

Group: Meteorites

Family: Stony meteorites

Texture: Crystalline

Structure: Massive

Composition: Olivene

Description: Found after meteorite showers

Rockname: Chert

Group: Sedimentary

Family: Cherts

Texture: Cryptocrystalline

Structure: Massive, bedded, laminated

Composition: Quartz

Description: Formed by chemical precipitation or accumulation of siliceous organic matter in ocean basins

Rockname: Chlorite schist

Group: Metamorphic

Family:

Texture: Crystalline, microcrystalline

Structure: Schistose

Composition: Chlorite, actinolite, mica

Description: Formed by low grade regional metamorphism of mainly pelitic sediments

Rockname: Chondrites

Group: Meteorites

Family: Stony meteorites

Texture: Crystalline, chondrules

Structure: Massive

Composition: Olivene, pyroxene, Fe-Ni alloy, troilite, chondrules (mineral granules)

Description: Found after meteorite showers

Rockname: Clinopyroxenite

Group: Igneous

Family: Pyroxenites

Texture: Crystalline

Structure: Massive

Composition: Augite (clinopyroxene)

Description: A type of ultramafic plutonic rock

Rockname: Coal

Group: Sedimentary

Family: Coals

Texture: Organic

Structure: Massive, bedded, laminated

Composition: Vitrinite, inertinite, liptonite, mineral matter

Description: Formed from plant remains that have undergone burial and maturation

Rockname: Comendite

Group: Igneous

Family: Rhyolites

Texture: Porphyritic

Structure: Massive, flow structure

Composition: Phenocrysts of quartz, K feldspar, aegerine, arfvedsonite or riebeckite, minor biotite

Description: Lava flows and intrusives of felsic composition

Rockname: Conglomerate

Group: Sedimentary

Family: Conglomerates

Texture: Clastic, grains > 4mm

Structure: Massive, bedded, laminated

Composition: Quartz, rock fragments, orthoclase (K feldspar) plagioclase, clay

Description: Deposited in rivers, alluvial fans and on scree slopes

Rockname: Coquina

Group: Sedimentary

Family: Limestones

Texture: Clastic

Structure: Massive, bedded, laminated

Composition: Coquina fossils

Description: A shelley limestone formed by the accumulation of coquina along shorelines

Rockname: Cordierite Hornfels

Group: Metamorphic

Family:

Texture: Microcrystalline

Structure: Massive

Composition: Cordierite, quartz, silliminite, kyanite

Description: Formed by contact metamorphism of sediments and igneous rocks

Rockname: Crystal Tuff (consolidated) or Crystal Ash
Group: Igneous
Family: Tuffs
Texture: Fragmental, grain size < 2mm
Structure: Massive, bedded, laminated
Composition: Crystal fragments predominant
Description: Ejecta accumulated by volcanic activity

Rockname: Dacite
Group: Igneous
Family: Dacites
Texture: Porphyritic
Structure: Massive, flow structure
Composition: Plagioclase, sanidine (K feldspar), quartz, biotite, hornblende (amphibole), pyroxenes
Description: Lava flows and intrusives of felsic composition

Rockname: Diabase
Group: Igneous
Family: Dolerites
Texture: Crystalline, fine grained
Structure: Massive
Composition: Plagioclase, hornblende (amphibole), pyroxenes, biotite, quartz
Description: Synonomous with Dolerite and Microgabbro

Rockname: Diatomite
Group: Sedimentary
Family:
Texture: Clastic
Structure: Bedded, laminated

Composition: Fossil diatoms
Description: Consisting entirely of fossil diatoms

Rockname: Diogenites
Group: Meteorites
Family: Stony meteorites
Texture: Crystalline
Structure: Massive
Composition: Hypersthene
Description: Found after meteorite showers

Rockname: Diorite
Group: Igneous
Family: Diorites
Texture: Crystalline
Structure: Massive
Composition: Plagioclase, pyroxenes, hornblende (amphibole), biotite, quartz, An < 50
Description: A mafic intrusive rock

Rockname: Dolerite
Group: Igneous
Family: Dolerites
Texture: Crystalline, fine grained
Structure: Massive
Composition: Plagioclase, pyroxenes
Description: Synonomous with Diabase and Microgabbro. Occurs as dykes

Rockname: Dolomite Carbonatite
Group: Igneous
Family: Carbonatites
Texture: Crystalline, microcyrstalline
Structure: Massive

Composition: Dolomite

Description: Carbonate rich (> 50%) igneous rocks

Rockname: Dolostone

Group: Sedimentary

Family: Limestones

Texture: Chemical

Structure: Massive, bedded, laminated

Composition: Dolomite, fossils, micrite, sparite, calcite

Description: A limestone consisting entirely of dolomite. Formed by replacement of calcite during or after deposition in ocean basins

Rockname: Dunite

Group: Igneous

Family: Peridotites

Texture: Crystalline

Structure: Massive

Composition: Olivene

Description: A type of ultramafic plutonic rock

Rockname: Dust Tuff

Group: Igneous

Family: Tuffs

Texture: Fragmental, consolidated, grain size < 2mm

Structure: Massive, bedded, laminated

Composition: Rock fragments, crystal fragments, glass fragments

Description: Ejecta accumulated by volcanic activity

Rockname: Eclogite

Group: Metamorphic

Family:

Texture: Crystalline

Structure: Massive
Composition: Omphacite (clinopyroxene), garnet
Description: A very high grade regional metamorphic rock

Rockname: Eclogite Facies
Group: Metamorphic
Family:
Texture: Crystalline
Structure: Massive
Composition: Diopside (clinopyroxene), garnet, orthopyroxene, olivene,
Description: A metamorphic facies containing very high grade regional metamorphic rocks of basic composition lacking plagioclase. The clinopyroxene has a high jadeite component (omphacitic).

Rockname: Enderbite
Group: Igneous, metamorphic
Family: Charnockites
Texture: Crystalline
Structure: Massive
Composition: Plagioclase, hypersthene, quartz, biotite, hornblende (amphibole)
Description: A granitoid containing hypersthene

Rockname: Enstatatite
Group: Igneous
Family: Pyroxenites
Texture: Crystalline
Structure: Massive
Composition: Enstatite
Description: A type of ultramafic plutonic rock

Rockname: Enstatite Chondrites
Group: Meteorites

Family: Stony meteorites

Texture: Crystalline, chondrules

Structure: Massive

Composition: Enstatite, Fe-Ni alloy, troilite, chondrules (mineral granules)

Description: Found after meteorite showers

Rockname: Epidote Amphibolite Facies

Group: Metamorphic

Family:

Texture: Crystalline

Structure: Massive, foliated, schistose

Composition: Epidote, plagioclase, hornblende, silliminite, andalusite, cordierite, cummingtonite, staurolite, biotite

Description: A metamorphic facies containing rocks that have undergone low to medium grade regional metamorphism

Rockname: Essexite

Group: Igneous

Family:

Texture: Crystalline

Structure: Massive

Composition: Augite, kaersutite and/or biotite, labradorite, minor K feldspar and nepheline

Description: Synonomous with Nepheline Monzogabbro or Monzodiorite

Rockname: Eucrites

Group: Meteorites

Family: Stony meteorites

Texture: Crystalline

Structure: Massive

Composition: Pyroxenes, plagioclase

Description: Found after meteorite showers

Rockname: Evaporite

Group: Sedimentary

Family: Evaporites

Texture: Chemical, nodular

Structure: Massive, bedded, laminated

Composition: Anhydrite, gypsum, halite, celestite, kainite, carnallite, sylvite, matrix

Description: A general term for rocks formed by precipitation during evaporation of brines

Rockname: Feldspathic Greywacke

Group: Sedimentary

Family: Sandstones

Texture: Clastic, grains 0.0625 to 2mm

Structure: Massive, bedded, laminated

Composition: Quartz, clay, orthoclase (K feldspar) plagioclase, clay 15% to 75%

Description: A feldspathic sandstone with > 15% clay matrix

Rockname: Fenite

Group: Igneous

Family:

Texture: Crystalline

Structure: Massive

Composition: K feldspar, sodic pyroxene, alkali amphibole or entirely K feldspar

Description: Usually occurs with carbonatites, Ijolites, Nepheline Syenites or peralkaline Granites

Rockname: Fergusite

Group: Igneous

Family: Foidites

Texture: Crystalline

Structure: Massive

Composition: (70% of K feldspar, nepheline, kalsitite and minor analcime), (30% of

pyroxene)
Description: Alkali plutonic rock

Rockname: Ferrocarbonatite
Group: Igneous
Family: Carbonatites
Texture: Crystalline, microcyrstalline
Structure: Massive
Composition: Siderite, ankerite
Description: Carbonate rich (> 50%) igneous rocks

Rockname: Fitzroyite
Group: Igneous
Family: Lamproites
Texture: Inequigranular, porphyritic
Structure: Massive
Composition: Phlogopite, leucite
Description: A type of lamproite

Rockname: Floatstone
Group: Sedimentary
Family: Limestones
Texture: Clastic, mud supported fabric, >10% grains >2mm
Structure: Massive, bedded
Composition: Fossils, clasts, micrite, sparite, calcite, dolomite, quartz
Description: A very coarse mud supported limestone

Rockname: Foid Diorite
Group: Igneous
Family:
Texture: Crystalline
Structure: Massive

Composition: Plagioclase, foids 10 - 60%, pyroxenes, hornblende (amphibole), biotite, An < 50

Description: Plutonic rocks of alkali composition

Rockname: Foid Gabbro

Group: Igneous

Family:

Texture: Crystalline

Structure: Massive

Composition: Plagioclase, foids 10 - 60%, pyroxenes, hornblende (amphibole), biotite, An > 50

Description: Plutonic rocks of alkali composition

Rockname: Foid Monzodiorite

Group: Igneous

Family:

Texture: Crystalline

Structure: Massive

Composition: Plagioclase , orthoclase, foids 10 - 60%, biotite, hornblende (amphibole), pyroxenes

Description: Plutonic rocks of alkali composition

Rockname: Foid Monzonite

Group: Igneous

Family: Monzonites

Texture: Crystalline

Structure: Massive

Composition: Orthoclase (K feldspar), plagioclase, biotite, hornblende (amphibole), foids < 10%

Description: Plutonic rocks of alkali composition

Rockname: Foid Monzosyenite

Group: Igneous

Family:

Texture: Crystalline

Structure: Massive

Composition: Orthoclase (K feldspar), plagioclase, foids 10 - 60%, biotite, hornblende (amphibole)

Description: Plutonic rocks of alkali composition

Rockname: Foid Syenite

Group: Igneous

Family:

Texture: Crystalline

Structure: Massive

Composition: Orthoclase (K feldspar), foids 10 - 60%, biotite, hornblende (amphibole), pyroxenes

Description: Plutonic rocks of alkali composition

Rockname: Foidites

Group: Igneous

Family: Foidites

Texture: Porphyritic

Structure: Massive, flow structure

Composition: Foids > 60%, biotite, hornblende (amphibole), pyroxenes, (minor plagioclase, sanidine (K feldspar))

Description: A volcanic rock of alkali composition occurring as lava flows and intrusions

Rockname: Foidolite

Group: Igneous

Family: Foidolites

Texture: Crystalline

Structure: Massive

Composition: Foids > 60%, pyroxenes, biotite, hornblende (amphibole), plagioclase, orthoclase

Description: Plutonic rocks of alkali composition

Rockname: Framestone

Group: Sedimentary

Family: Limestones

Texture: Biogenic

Structure: Massive, bedded, laminated

Composition: Fossils, algae, micrite, sparite, calcite

Description: A type of boundstone with organisms forming the framework

Rockname: Gabbro

Group: Igneous

Family: Gabbros

Texture: Crystalline

Structure: Massive

Composition: Plagioclase, pyroxenes, hornblende (amphibole), biotite, An > 50

Description: A mafic plutonic rock

Rockname: Gabbronorite

Group: Igneous

Family: Gabbros

Texture: Crystalline

Structure: Massive

Composition: Augite (clinopyroxene), enstatite (orthopyroxene), hypersthene, plagioclase

Description: A mafic plutonic rock

Rockname: Gaussbergite

Group: Igneous

Family: Lamproites

Texture: Inequigranular, porphyritic

Structure: Massive

Composition: Olivene, glass, leucite

Description: A type of lamproite

Rockname: Geode

Group: Igneous

Family: Agates

Texture: Cryptocrystalline

Structure: Banded with interior crystal filled cavity

Composition: Quartz

Description: Formed by shrinkage or partial deposition of silica gels in cavities of volcanic rocks.

Rockname: Glaucophane Schist

Group: Metamorphic

Family:

Texture: Crystalline

Structure: Schistose, foliated

Composition: Glaucophane, jadeite, lawsonite

Description: High pressure regional metamorphic rocks of basic composition and containing glaucophane

Rockname: Glaucophane Schist Facies

Group: Metamorphic

Family:

Texture: Crystalline

Structure: Schistose, foliated

Composition: Glaucophane, jadeite, lawsonite

Description: A metamorphic facies containing high pressure regional metamorphic rocks usually of basic composition and which contain glaucophane and/or lawsonite

Rockname: Gneiss

Group: Metamorphic

Family:

Texture: Crystalline

Structure: Foliated, banded

Composition: Plagioclase, quartz, orthoclase (K feldspar), hornblende, biotite, muscovite, silliminite

Description: Formed by medium to high grade regional metamorphism of quartzo-feldspathic rocks.

Rockname: Gossan

Group: Duricrust

Family:

Texture: Chemical, weathered

Structure: Boxwork, skeletal, network

Composition: Iron oxides, relict ore minerals

Description: Weathered product of orebody showing relict ore mineral textures

Rockname: Grainstone

Group: Sedimentary

Family: Limestones

Texture: Clastic, grain supported fabric

Structure: Massive, bedded

Composition: Fossils, clasts, sparite, calcite, dolomite, quartz

Description: A grain supported limestone without a micritic matrix

Rockname: Granite

Group: Igneous

Family: Granites

Texture: Crystalline

Structure: Massive

Composition: Quartz, orthoclase (K feldspar), plagioclase, biotite, hornblende (amphibole)

Description: A very common plutonic rock of felsic composition

Rockname: Granodiorite

Group: Igneous

Family: Granites

Texture: Crystalline

Structure: Massive

Composition: Quartz, plagioclase, orthoclase (K feldspar), biotite, hornblende (amphibole)

Description: A variety of granite

Rockname: Granulite

Group: Metamorphic

Family:

Texture: Crystalline, granulitic

Structure: Banded, foliated, streaked, massive

Composition: Plagioclase, garnet, silliminite, diopside, kyanite, quartz, biotite

Description: A high grade metamorphic rock with a granular (granulitic) texture

Rockname: Granulite Facies

Group: Metamorphic

Family:

Texture: Crystalline

Structure: Banded, foliated, massive

Composition: Plagioclase, garnet, silliminite, diopside, kyanite, quartz, biotite

Description: A metamorphic facies containing rocks formed by high grade regional metamorphism

Rockname: Greenschist

Group: Metamorphic

Family:

Texture: Crystalline, recrystallized

Structure: Schistose

Composition: Actinolite, chlorite, plagioclase, muscovite mica, biotite, stilpnomelane

Description: A general term for schistose rocks formed by low grade regional metamorphism

Rockname: Greenschist Facies

Group: Metamorphic

Family:

Texture: Crystalline, recrystallized

Structure: Massive, foliated, schistose

Composition: Actinolite, chlorite, plagioclase, muscovite mica, biotite, stilpnomelane

Description: A metamorphic facies containing rocks that have undergone low grade regional metamorphism

Rockname: Greenstone

Group: Metamorphic

Family:

Texture: Crystalline, recrystallized

Structure: Massive

Composition: Chlorite, actinolite, serpentine, plagioclase, hornblende, orthopyroxene, clinopyroxene

Description: An old field term for metamorphosed basalts

Rockname: Greywacke

Group: Sedimentary

Family: Sandstones

Texture: Clastic, grains 0.0625 to 2mm

Structure: Massive, bedded, laminated

Composition: Quartz, clay, orthoclase (K feldspar) plagioclase, rock fragments, clay 15% to 75%

Description: A sandstone with > 15% clay matrix

Rockname: Gypsum Rock

Group: Sedimentary

Family: Evaporites

Texture: Chemical, nodular

Structure: Massive, bedded, laminated

Composition: Gypsum

Description: Formed by precipitation during evaporation of brines

Rockname: Halite Rock

Group: Sedimentary

Family: Evaporites

Texture: Chemical

Structure: Massive, bedded, laminated

Composition: Halite

Description: Formed by precipitation during evaporation of brines

Rockname: Harzburgite

Group: Igneous

Family: Peridotites

Texture: Crystalline

Structure: Massive

Composition: Olivene, enstatite (orthopyroxene), hypersthene

Description: A type of ultramafic plutonic rock

Rockname: Hauynite

Group: Igneous

Family: Foidites

Texture: Porphyritic

Structure: Massive, flow structure

Composition: Hauyne, augite (clinopyroxene), olivene, glass, plagioclase

Description: A volcanic rock of alkali composition occurring as lava flows and intrusions

Rockname: Hawaiite

Group: Igneous

Family: Basalts

Texture: Porphyritic

Structure: Massive, flow structure

Composition: Plagioclase, augite (clinopyroxene), hornblende, minor olivene, glass

Description: A variety of Trachybasalt defined chemically with $Na_2O - 2.0 > K_2O$

Rockname: Hexahedrites

Group: Meteorites

Family: Iron meteorites

Texture: Crystalline, hexahedron crystals

Structure: Massive

Composition: Kamacite, troilite, graphite

Description: With hexahedron (cubic) crystals of kamacite.

Rockname: Hornblende Gabbro

Group: Igneous

Family: Gabbros

Texture: Crystalline

Structure: Massive

Composition: Hornblende, plagioclase

Description: A mafic plutonic rock

Rockname: Hornblende Hornfels Facies

Group: Metamorphic

Family:

Texture: Microcrystalline

Structure: Massive

Composition: Plagioclase, hornblende, silliminite, andalusite, cordierite, cummingtonite, kyanite, staurolite, biotite

Description: A metamorphic facies containing rocks that have undergone medium grade contact metamorphism

Rockname: Hornblende Peridotite
Group: Igneous
Family: Peridotites
Texture: Crystalline
Structure: Massive
Composition: Olivene, hornblende (amphibole)
Description: A type of ultramafic plutonic rock

Rockname: Hornblende Pyroxenites
Group: Igneous
Family: Pyroxenites
Texture: Crystalline
Structure: Massive
Composition: Hornblende (amphibole), augite (clinopyroxene), enstatite (orthopyroxene), hypersthene
Description: A type of ultramafic plutonic rock

Rockname: Hornblendite
Group: Igneous
Family: Gabbros
Texture: Crystalline
Structure: Massive
Composition: Hornblende (amphibole), plagioclase
Description: A type of ultramafic plutonic rock

Rockname: Hornfels
Group: Metamorphic
Family:
Texture: Microcrystalline

Structure: Massive
Composition: Cordierite, quartz, silliminite, kyanite, almandine garnet
Description: Formed by contact metamorphism of sediments and igneous rocks

Rockname: Hortschiefer
Group: Metamorphic
Family:
Texture: Fragmental, recystallized
Structure: Massive, foliated
Composition: Quartz, plagioclase, orthoclase (K feldspar), garnet, biotite, chlorite, epidote, calcite, crushed matrix, rock fragments, mineral fragments
Description: Composed of fragmental material in a crushed matrix that has undergone minor recrystallization

Rockname: Howardites
Group: Meteorites
Family: Stony meteorites
Texture: Crystalline
Structure: Massive
Composition: Pyroxenes, plagioclase
Description: Found after meteorite showers

Rockname: Hyalomylonite
Group: Metamorphic
Family:
Texture: Fragmental, glassy matrix
Structure: Massive, flow
Composition: Quartz, plagioclase, orthoclase (K feldspar), garnet, biotite, chlorite, epidote, calcite in a glassy matrix
Description: Formed at high temperatures and pressures along fault and shear zones. Variable fragmental material in a glassy matrix. Synonomous with pseudotachylite

Rockname: Hypersthene Alkali Feldspar Syenite

Group: Igneous, metamorphic

Family: Charnockites

Texture: Crystalline

Structure: Massive

Composition: Orthoclase (K feldspar), hypersthene, biotite, hornblende (amphibole)

Description: A granitoid containing hypersthene

Rockname: Hypersthene Syenite

Group: Igneous, metamorphic

Family: Charnockites

Texture: Crystalline

Structure: Massive

Composition: Orthoclase (K feldspar), plagioclase, hypersthene, biotite, hornblende (amphibole)

Description: A granitoid containing hypersthene

Rockname: Ignimbrite

Group: Igneous

Family: Tuffs

Texture: Fragmental, welded

Structure: Massive, bedded, laminated

Composition: Rock fragments, crystal fragments, glass fragments

Description: A welded tuff containing crystal and rock fragments in a matrix of glass shards

Rockname: Ijolite

Group: Igneous

Family: Foidolites

Texture: Crystalline

Structure: Massive

Composition: Pyroxene, 30 to 70% nepheline

Description: A mesocratic variety of Nephelinolite (Foidolite)

Rockname: Iodranites

Group: Meteorites

Family: Stony iron meteorites

Texture: Crystalline

Structure: Massive

Composition: Fe-Ni alloys, bronzite, olivene

Description: Found after meteorite showers

Rockname: Ironstone

Group: Sedimentary

Family: Ironstones

Texture: Chemical, pisolitic, cryptocystalline, oolitic

Structure: Massive, bedded, laminated

Composition: Hematite, magnetite, limonite, siderite, chamosite

Description: A general term for rocks chiefly containing iron oxide minerals

Rockname: Italite

Group: Igneous

Family: Foidites

Texture: Porphyritic

Structure: Massive, flow

Composition: Leucite, minor mafics

Description: A leucocratic variety of Leucitite

Rockname: Jotunite

Group: Igneous, metamorphic

Family: Charnockites

Texture: Crystalline

Structure: Massive

Composition: Plagioclase, orthoclase (K feldspar), hypersthene, biotite, hornblende

(amphibole)

Description: A granitoid containing hypersthene

Rockname: Jumillite

Group: Igneous

Family: Lamproites

Texture: Inequigranular, porphyritic

Structure: Massive

Composition: Richterite, diopside, olivene, phlogopite groundmass (madupitic)

Description: A type of lamproite

Rockname: Kersantite

Group: Igneous

Family: Lamprophyres

Texture: Porphyritic

Structure: Massive

Composition: Biotite, augite, plagioclase, olivene, serpentine (sec), chlorite (sec), plag > or

Description: A mafic to ultramafic minor intrusive rock

Rockname: Kimberlite

Group: Igneous

Family: Kimberlites

Texture: Inequigranular

Structure: Massive

Composition: Olivene, phlogopite, calcite, diopside (clinopyroxene), spinel, garnet, enstatite (orthopyroxene), serpentine (sec), xenocrysts, xenoliths

Description: A general term for highly potassic ultramafic rocks with abundant megacrysts, xenocrysts and xenoliths. Occurs as pipes. The major source of diamonds.

Rockname: Komatiite

Group: Igneous

Family:

Texture: Porphyritic, spinifex

Structure: Massive, flow structure

Composition: Olivene, augite (clinopyroxene), glass

Description: A high magnesium ultramafic lava usually with a well developed spinifex texture

Rockname: Kugdite or Olivene Melilitolite

Group: Igneous

Family: Melilites

Texture: Crystalline

Structure: Massive

Composition: Melilite, olivene

Description:

Rockname: Lamproite

Group: Igneous

Family: Lamproites

Texture: Inequigranular, porphyritic

Structure: Massive

Composition: Phlogopite, diopside (clinopyroxene), richterite, enstatite (orthopyroxene), Sanidine (K feldspar), leucite, glass, serpentine (sec), xenocrysts

Description: A general term for highly potassic mafic to ultramafic rocks occurring as small extrusions, dykes and pipes.

Rockname: Lamprophyre

Group: Igneous

Family: Lamprophyres

Texture: Porphyritic, inequigranular

Structure: Massive

Composition: Olivene, diopside, augite (clinopyroxene), enstatite (orthopyroxene) , biotite, hornblende (amphibole), melilite, nepheline (foids), plagiocalse feldspar, orthoclase, xenocrysts

Description: A general term for a group of mafic to ultramafic minor intrusive rocks with feldspars and foids restricted to the groundmass

Rockname: Lapilli tephra
Group: Igneous
Family: Tephra
Texture: Fragmental, unconsolidated, grain size > 2mm and < 64mm
Structure: Massive, bedded, laminated
Composition: Rock fragments, crystal fragments, glass fragments
Description: Ejecta accumulated by volcanic activity

Rockname: Lapilli Tuff
Group: Igneous
Family: Tuffs
Texture: Fragmental, consolidated, grain size > 2mm and < 64mm
Structure: Massive, bedded, laminated
Composition: Rock fragments, crystal fragments, glass fragments
Description: Ejecta accumulated by volcanic activity

Rockname: Larvikite
Group: Igneous
Family: Syenites
Texture: Crystalline
Structure: Massive
Composition: Rhomb shaped feldspars, schiller structure, barkevikite, augite, lepidomelane, plus/minus olivene, nepheline or quartz
Description:

Rockname: Laterite
Group: Sedimentary
Family: Ironstones
Texture: Chemical, pisolitic

Structure: Massive

Composition: Limonite, goethite, iron oxides

Description: Formed by precipitation of iron oxides in the surface weathering profile

Rockname: Latite

Group: Igneous

Family: Latites

Texture: Porphyritic

Structure: Massive, flow structure

Composition: Plagioclase, sanidine (K feldspar), biotite, hornblende (amphibole), pyroxenes

Description: Lava flows and intrusives of felsic composition

Rockname: Lawsonite Schist

Group: Metamorphic

Family:

Texture: Crystalline

Structure: Schistose, foliated

Composition: Lawsonite, jadeite

Description: High pressure regional metamorphic rocks of basic composition and containing lawsonite

Rockname: Lherzolite

Group: Igneous

Family: Peridotites

Texture: Crystalline

Structure: Massive

Composition: Olivene, enstatite (orthopyroxene), hypersthene, hypersthene, augite (clinopyroxene)

Description: A type of ultramafic plutonic rock

Rockname: Lignite or Brown Coal

Group: Sedimentary

Family: Coals
Texture: Organic
Structure: Massive, bedded, laminated
Composition: Vitrinite, inertinite, liptonite, mineral matter
Description: A low grade (low carbon content, high volatile content) coal

Rockname: Limburgite
Group: Igneous
Family: Basanites
Texture: Porphyritic
Structure: Massive, flow
Composition: Pyroxene, olivene, opaques in groundmass of same with glass, no feldspars
Description: Alkali volcanic rock. A variety of Basanite containing glass

Rockname: Limestone
Group: Sedimentary
Family: Limestones
Texture: Clastic, biogenic, chemical, pelletal, oolitic
Structure: Massive, bedded, laminated
Composition: Fossils, clasts, micrite, sparite, calcite, dolomite, quartz
Description: A general term for carbonate rocks formed by chemical precipitation, mechanical accumulation or biological formation of calcium carbonate

Rockname: Lipartite
Group: Igneous
Family: Rhyolites
Texture: Porphyritic
Structure: Massive, flow structure
Composition: Sanidine (K feldspar), quartz, plagioclase, biotite, hornblende (amphibole), pyroxenes
Description: Synonomous with Rhyolite

Rockname: Litharenite

Group: Sedimentary

Family: Sandstones

Texture: Clastic, grains 0.0625 to 2mm

Structure: Massive, bedded, laminated

Composition: Quartz, rock fragments, clay, clay < 15%

Description: A lithic sandstone with < 15% clay matrix

Rockname: Lithic Arkose

Group: Sedimentary

Family: Sandstones

Texture: Clastic, grains 0.0625 to 2mm

Structure: Massive, bedded, laminated

Composition: Quartz, orthoclase (K feldspar), plagioclase, minor rock fragments, clay < 15%

Description: A feldspathic and lithic sandstone with < 15% clay matrix

Rockname: Lithic Greywacke

Group: Sedimentary

Family: Sandstones

Texture: Clastic, grains 0.0625 to 2mm

Structure: Massive, bedded, laminated

Composition: Quartz, clay, rock fragments, clay 15% to 75%

Description: A lithic sandstone with > 15% clay matrix

Rockname: Lithic Tuff (consolidated) or Lithic Ash

Group: Igneous

Family: Tuffs

Texture: Fragmental, grain size < 2mm

Structure: Massive, bedded, laminated

Composition: Rock fragments predominant

Description: Ejecta accumulated by volcanic activity

Rockname: Madupite

Group: Igneous

Family: Lamproites

Texture: Inequigranular, porphyritic, madupitic

Structure: Massive

Composition: Diopside, xenocrysts, phlogopite groundmass (madupitic)

Description: A type of lamproite

Rockname: Malignite

Group: Igneous

Family: Syenites

Texture: Crystalline

Structure: Massive

Composition: Aegerine augite, (orthoclase and nepheline in equal amounts), mafic minerals

Description: A mesocratic variety of Syenite, rich in aegerine augite

Rockname: Mamillite

Group: Igneous

Family: Lamproites

Texture: Inequigranular, porphyritic

Structure: Massive

Composition: Richterite, leucite

Description: A type of lamproite

Rockname: Mangerite

Group: Igneous, metamorphic

Family: Charnockites

Texture: Crystalline

Structure: Massive

Composition: Orthoclase (K feldspar), hypersthene, plagioclase, biotite, hornblende

(amphibole)

Description: A granitoid containing hypersthene

Rockname: Marble

Group: Metamorphic

Family:

Texture: Crystalline

Structure: Massive

Composition: Calcite, dolomite, brucite

Description: Formed by high temperature contact metamorphism of limestone

Rockname: Meimechite

Group: Igneous

Family:

Texture: Porphyritic

Structure: Massive, flow structure

Composition: Olivene phenocrysts in a groundmass of olivene, augite (clinopyroxene) and glass

Description: A high magnesium ultramafic lava

Rockname: Melilite

Group: Igneous

Family: Melilites

Texture: Porphyritic

Structure: Massive

Composition: Melilite, plus/minus augite (clinopyroxene)

Description:

Rockname: Melilitolite

Group: Igneous

Family: Melilites

Texture: Crystalline

Structure: Massive
Composition: Melilite
Description:

Rockname: Melteigite
Group: Igneous
Family: Foidolites
Texture: Crystalline
Structure: Massive
Composition: 10 - 30% nepheline, mafic minerals
Description: A melanocratic variety of Nephelinolite (foidolite)

Rockname: Mesosiderites
Group: Meteorites
Family: Stony iron meteorites
Texture: Crystalline
Structure: Massive
Composition: Fe-Ni alloys, pyroxene, plagioclase
Description: Found after meteorite showers

Rockname: Meta-Basalt
Group: Metamorphic
Family:
Texture: Crystalline, recrystallized
Structure: Massive
Composition: Chlorite, actinolite, serpentine, plagioclase, hornblende
Description: Slightly metamorphosed basalts retaining their original texture. Synonomous with the old field name greenstone

Rockname: Meta-Basite
Group: Metamorphic
Family:

Texture: Crystalline, recrystallized

Structure: Massive, foliated

Composition: Plagioclase, hornblende, diopside, orthopyroxene, clinopyroxene, chlorite, actinolite

Description: A general term for metamorphosed igneous rocks of basic composition.

Rockname: Meta-Pelite

Group: Metamorphic

Family:

Texture: Recrystallized, microcrystalline

Structure: Massive, foliated, schistose

Composition: Biotite, muscovite, actinolite, chlorite, quartz, feldspar, clay

Description: A general term for metamorphosed argillaceous sediments

Rockname: Miaskite

Group: Igneous

Family:

Texture: Crystalline

Structure: Massive

Composition: Perthitic orthoclase (K feldspar), oligoclase (plagioclase), nepheline, biotite

Description: A leucocratic variety of Biotite, Nepheline Monzosyenite

Rockname: Mica Schist

Group: Metamorphic

Family:

Texture: Crystalline, microcrystalline

Structure: Schistose

Composition: Biotite, muscovite, actinolite, chlorite

Description: Formed by low grade regional metamorphism of mainly pelitic sediments

Rockname: Microgabbro

Group: Igneous

Family: Gabbros
Texture: Crystalline, fine grained
Structure: Massive
Composition: Plagioclase, pyroxenes, hornblende (amphibole), biotite, quartz
Description: A variety of gabbro

Rockname: Microgranite
Group: Igneous
Family: Granites
Texture: Microcrystalline
Structure: Massive
Composition: Quartz, orthoclase (K feldspar), plagioclase, biotite, hornblende (amphibole)
Description: A variety of granite

Rockname: Migmatite
Group: Metamorphic
Family:
Texture: Crystalline
Structure: Various types of banded and foliated structures
Composition: Plagioclase, quartz, orthoclase (K feldspar), hornblende, biotite, muscovite, silliminite
Description: A heterogenous rock formed by medium to high grade regional metamorphism of quatzo-feldspathic igneous and sedimentary rocks.

Rockname: Minette, Igneous
Group: Igneous
Family: Lamprophyres
Texture: Porphyritic
Structure: Massive
Composition: Biotite, augite, orthoclase, olivene, serpentine (sec), chlorite (sec), or > plag
Description: A mafic to ultramafic minor intrusive rock

Rockname: Minette, Sedimentary

Group: Sedimentary

Family: Ironstones

Texture: Chemical, clastic, oolitic

Structure: Massive, bedded, laminated

Composition: Chamosite, siderite, limonite, fossils, micrite

Description: Deposited in shallow marine shoals

Rockname: Missourite

Group: Igneous

Family: Foidolites

Texture: Crystalline

Structure: Massive

Composition: Clinopyroxene, leucite, olivene

Description: A melanocratic variety of Foidolite

Rockname: Monchiquite

Group: Igneous

Family: Lamprophyres

Texture: Porphyritic

Structure: Massive

Composition: Barkevite, kaersutite (amphibole), augite, olivene, biotite, glass or foids

Description: A mafic to ultramafic minor intrusive rock

Rockname: Monzodiorite

Group: Igneous

Family: Granites

Texture: Crystalline

Structure: Massive

Composition: Plagioclase, orthoclase (K feldspar), quartz, pyroxenes, biotite, hornblende (amphibole)

Description: Intermediate in composition between monzonite and diorite

Rockname: Monzonite
Group: Igneous
Family: Syenites
Texture: Crystalline
Structure: Massive
Composition: Plagioclase, K- feldspar, biotite, hornblende (amphibole), pyroxenes
Description: Dominant felsic mineral is K-feldspar. Lacks quartz

Rockname: Mudstone
Group: Sedimentary
Family: Mudstones
Texture: Clastic, grains < 0.0039mm
Structure: Massive, bedded, laminated
Composition: Clay, quartz,   glauconite, pyrite
Description: Deposited in seas, lakes and rivers

Rockname: Mudstone, Carbonate
Group: Sedimentary
Family: Limestones
Texture: Clastic
Structure: Massive, bedded, laminated
Composition: Micrite, <10% grains
Description: A limestone consisting of carbonate mud

Rockname: Mugearite
Group: Igneous
Family: Basalts
Texture: Porphyritic
Structure: Massive, flow structure
Composition: Phenocrysts of olivene, augite and magnetite in a groundmass of plagioclase, augite, magnetite and K feldspar

Introduction to Geology

Description: A variety of Basaltic Trachyandesite defined chemically with Na2O - 2.0 >= K2O

Rockname: Mylonite
Group: Metamorphic
Family:
Texture: Fragmental
Structure: Foliated
Composition: Crushed matrix, rock fragments, mineral fragments
Description: Composed of fragmental material in a crushed matrix formed by medium to high grade pressure metamorphism

Rockname: Nakhlites
Group: Meteorites
Family: Stony meteorites
Texture: Crystalline
Structure: Massive
Composition: Diopside, olivene
Description: Found after meteorite showers

Rockname: Natrocarbonatite
Group: Igneous
Family: Carbonatites
Texture: Crystalline, microcyrstalline
Structure: Massive
Composition: K, Na and Ca carbonates
Description: Carbonate rich (> 50%) igneous rocks

Rockname: Nephelinite
Group: Igneous
Family: Foidites
Texture: Porphyritic

Structure: Massive, flow structure

Composition: Nepheline, augite (clinopyroxene), glass

Description: A volcanic rock of alkali composition occurring as lava flows and intrusions

Rockname: Nephelinolite

Group: Igneous

Family: Foidolites

Texture: Crystalline

Structure: Massive

Composition: Nepheline, mafic minerals

Description: A variety of foidolite further subdivided into Urtite, Ijolite and Melteigite based on mafic mineral content

Rockname: Norite

Group: Igneous

Family: Gabbros

Texture: Crystalline

Structure: Massive

Composition: Plagioclase, hypersthene (orthopyroxene)

Description: A mafic plutonic rock

Rockname: Noseanite

Group: Igneous

Family: Foidites

Texture: Porphyritic

Structure: Massive, flow

Composition: Nosean, amphibole, nepheline

Description: A variety of nephelinite containing nosean, nepheline and amphibole

Rockname: Obsidian

Group: Igneous

Family:

Texture: Amorphous

Structure: Massive

Composition: Glass

Description: A variety of volcanic glass usually black or brown

Rockname: Octahedrites

Group: Meteorites

Family: Iron meteorites

Texture: Crystalline, octahedron texture

Structure: Widmanstatten structure obvious after etching

Composition: Kamacite, taenite, plessite, graphite

Description: Named after the octahedron texture called widmanstatten structure.

Rockname: Oil Shale

Group: Sedimentary

Family: Carbonaceous

Texture: Organic, clastic

Structure: Laminated, bedded

Composition: Clay, algae and fungae

Description: Rich in kerogen but do not contain free oil

Rockname: Olivene Basalt

Group: Igneous

Family: Basalts

Texture: Porphyritic

Structure: Massive, flow structure

Composition: Olivene, plagioclase, augite (clinopyroxene), glass

Description: Olivene bearing Basalt

Rockname: Olivene Gabbronorite

Group: Igneous

Family: Gabbros

Texture: Crystalline

Structure: Massive

Composition: Plagioclase, augite (clinopyroxene), enstatite (orthopyroxene), olivene, hypersthene

Description: A mafic plutonic rock

Rockname: Olivene Hornblende Pyroxenite

Group: Igneous

Family: Pyroxenites

Texture: Crystalline

Structure: Massive

Composition: Augite (clinopyroxene), enstatite (orthopyroxene), hypersthene, hornblende (amphibole), olivene

Description: A type of ultramafic plutonic rock

Rockname: Olivene Melilite

Group: Igneous

Family: Melilites

Texture: Porphyritic

Structure: Massive

Composition: Melilite, augite (clinopyroxene), olivene

Description:

Rockname: Olivene Orthopyroxenite

Group: Igneous

Family: Pyroxenites

Texture: Crystalline

Structure: Massive

Composition: Enstatite, hypersthene (orthopyroxene), olivene

Description: A type of ultramafic plutonic rock

Rockname: Olivene Pyroxene Hornblendite

Group: Igneous

Family: Pyroxenites

Texture: Crystalline

Structure: Massive

Composition: Hornblende (amphibole), augite (clinopyroxene), enstatite (orthopyroxene), hypersthene, olivene

Description: A type of ultramafic plutonic rock

Rockname: Olivene Pyroxenite

Group: Igneous

Family: Pyroxenites

Texture: Crystalline

Structure: Massive

Composition: Olivene, augite (clinopyroxene)

Description: A type of ultramafic plutonic rock

Rockname: Olivene Tholeiite

Group: Igneous

Family: Basalts

Texture: Porphyritic

Structure: Massive, flow structure

Composition: Plagioclase (laboradorite), pigeonite, olivene, hypersthene, augite (clinopyroxene), glass

Description: A type of Tholeiitic Basalt containing olivene.

Rockname: Olivene Uncompahgrite or Olivene Pyroxene Melilitolite

Group: Igneous

Family: Melilites

Texture: Crystalline

Structure: Massive

Composition: Melilite, olivene, augite (clinopyroxene)

Description:

Rockname: Olivene Websterite

Group: Igneous

Family: Pyroxenites

Texture: Crystalline

Structure: Massive

Composition: orthopyroxene, augite (clinopyroxene), olivene

Description: A type of ultramafic plutonic rock

Rockname: Opdalite

Group: Igneous, metamorphic

Family: Charnockites

Texture: Crystalline

Structure: Massive

Composition: Plagioclase, orthoclase (K feldspar), quartz, hypersthene, biotite, hornblende (amphibole)

Description: A granitoid containing hypersthene

Rockname: Orendite

Group: Igneous

Family: Lamproites

Texture: Inequigranular, porphyritic

Structure: Massive

Composition: Phlogopite, sanidine (K feldspar), diopside

Description: A type of lamproite

Rockname: Orthoconglomerate

Group: Sedimentary

Family: Conglomerates

Texture: Clastic, grains > 4mm

Structure: Massive, bedded, laminated

Composition: Quartz, rock fragments, orthoclase (K feldspar) plagioclase, clay, matrix < 15%

Description: Deposited in rivers, alluvial fans and on scree slopes

Rockname: Orthopyroxenite

Group: Igneous

Family: Pyroxenites

Texture: Crystalline

Structure: Massive

Composition: Enstatite, hypersthene (orthopyroxene)

Description: A type of ultramafic plutonic rock

Rockname: Packstone

Group: Sedimentary

Family: Limestones

Texture: Clastic, grain supported fabric

Structure: Massive, bedded

Composition: Fossils, clasts, micrite, sparite, calcite, dolomite, quartz, >10% grains

Description: A grain supported limestone with a micritic matrix

Rockname: Pallasites

Group: Meteorites

Family: Stony iron meteorites

Texture: Crystalline

Structure: Massive

Composition: Fe-Ni alloys, olivene

Description: Found after meteorite showers

Rockname: Paraconglomerate or Diamictite

Group: Sedimentary

Family: Conglomerates

Texture: Clastic, grains > 4mm

Structure: Massive, bedded, laminated

Composition: Quartz, rock fragments, orthoclase (K feldspar) plagioclase, clay, matrix > 15%

Description: Deposited in rivers, alluvial fans and on scree slopes

Rockname: Peat

Group: Sedimentary

Family: Coals

Texture: Organic

Structure: Massive, bedded, laminated

Composition: Decayed vegetable matter

Description: Forms in swamps and bogs

Rockname: Pegmatite

Group: Igneous

Family:

Texture: Crystalline, coarse grained

Structure: Massive

Composition: Various

Description: Macrocrystalline variety of igneous rocks comimon as veins and dykes often rich in gems

Rockname: Pelite

Group: Metamorphic

Family:

Texture: Recystallized

Structure: Massive, laminated, bedded

Composition: Illite (clay), prehnite, pumpellyite, chlorite, muscovite, biotite, quartz

Description: A low grade metamorphic rock generally derived from metamorphism of mudstones claystones

Rockname: Phonolite

Group: Igneous

Family: Phonolites

Texture: Porphyritic

Structure: Massive, flow structure

Composition: Sanidine (K feldspar), foids

Description: Alkali volcanic rock consisting primarily of sanidine and foids

Rockname: Phonolitic Tephrite

Group: Igneous

Family: Tephrites

Texture: Porphyritic

Structure: Massive, flow structure

Composition: Plagioclase, sanidine (K feldspar), foids, clinopyroxene

Description: A volcanic rock of alkali composition occurring as lava flows and intrusions

Rockname: Phosphorite

Group: Sedimentary

Family: Phosphorites

Texture: Chemical, Clastic

Structure: Massive, bedded, laminated

Composition: Apatite, phosphates

Description: Formed by precipitation from sea water, replacement of limestones or shelly marine sediments, in guano deposits and accumulation of grains from pre-existing rocks

Rockname: Phyllite

Group: Metamorphic

Family:

Texture: Recrystallized

Structure: Foliated

Composition: Clay, quartz, orthoclase (K feldspar), plagioclase, muscovite, chlorite, biotite

Description: Formed during low pressure regional metamorphism of argillaceous rocks

Rockname: Picrite

Group: Igneous

Family:

Texture: Porphyritic

Structure: Massive, flow structure

Composition: Olivene, augite (clinopyroxene), plagioclase

Description: A type of ultramafic volcanic rock

Rockname: Picrobasalt

Group: Igneous

Family: Basalts

Texture: Porphyritic

Structure: Massive, flow structure

Composition: Augite (clinopyroxene), olivene, orthopyroxene, plagioclase

Description: A volcanic rock of ultrabasic composition

Rockname: Plagioclasite

Group: Igneous

Family: Anorthosites

Texture: Crystalline

Structure: Massive

Composition: Plagioclase

Description: Synonomous with Anorthosite

Rockname: Plagiogranite (Plagioclase Granite)

Group: Igneous

Family: Granites

Texture: Crystalline

Structure: Massive

Composition: Plagioclase, quartz (<10% of biotite and hornblende)

Description: A plagioclase rich granite. Synonomous with Trondhjemite anmd leucocratic Tonalite

Rockname: Polzenite
Group: Igneous
Family: Lamprophyres
Texture: Porphyritic
Structure: Massive
Composition: Melilite, biotite, augite, olivene, calcite, glass or foids
Description: A mafic to ultramafic minor intrusive rock

Rockname: Porphyry
Group: Igneous
Family:
Texture: Porphyritic, phenocrysts in a fine grained groundmass
Structure: Massive
Composition: Various
Description: A general term for igneous rocks containing phenocrysts in a fine grained groundmass. Occurs as dykes, intrusions and extrusions

Rockname: Prehnite Pumpellyite Facies
Group: Metamorphic
Family:
Texture: Recrystallized
Structure: Massive
Composition: Clay, prehnite, pumpellyite, chlorite, epidote, plagioclase
Description: A metamorphic facies containing very low grade metamorphic rocks that have undergone burial metamorphism

Rockname: Psammite
Group: Metamorphic
Family:

Texture: Recrystallized, medium to coarse grained

Structure: Massive, foliated

Composition: Biotite, muscovite, actinolite, chlorite, quartz, feldspar, chlorite

Description: A general term for metamorphosed arenaceous sediments

Rockname: Pseudotachylite

Group: Metamorphic

Family:

Texture: Glassy matrix, fragmental

Structure: Massive, flow

Composition: Quartz, plagioclase, orthoclase (K feldspar), garnet, biotite, chlorite, epidote, calcite with a glassy matrix

Description: Formed at high temperatures and pressures along fault and shear zones. Variable fragmental material in a glassy matrix. Synonomous with hyalomylonite

Rockname: Pyroclastic Breccia or Agglomerate

Group: Igneous

Family: Pyroclastic

Texture: Fragmental, consolidated, grain size > 64mm

Structure: Massive, bedded

Composition: Rock fragments, crystal fragments, glass fragments

Description: Ejecta accumulated by volcanic activity

Rockname: Pyroxene Hornblende Gabbronorite

Group: Igneous

Family: Gabbros

Texture: Crystalline

Structure: Massive

Composition: Augite (clinopyroxene), enstatite (orthopyroxene), plagioclase, hypersthene, hornblende (amphibole), plagioclase

Description: A mafic plutonic rock

Rockname: Pyroxene Hornblende Peridotite

Group: Igneous

Family: Peridotites

Texture: Crystalline

Structure: Massive

Composition: Olivene, augite (clinopyroxene), enstatite (orthopyroxene), hypersthene, hornblende (amphibole)

Description: A type of ultramafic plutonic rock

Rockname: Pyroxene Hornblendite

Group: Igneous

Family: Pyroxenites

Texture: Crystalline

Structure: Massive

Composition: Augite (clinopyroxene), enstatite (orthopyroxene), hypersthene, hornblende (amphibole)

Description: A type of ultramafic plutonic rock

Rockname: Pyroxene Hornfels

Group: Metamorphic

Family:

Texture: Microcrystalline

Structure: Massive

Composition: Diopside (clinopyroxene), plagioclase, almandine garnet, vesuvianite

Description: Formed by high temperature contact metamorphism of igneous rocks

Rockname: Pyroxene Hornfels Facies

Group: Metamorphic

Family:

Texture: Microcrystalline

Structure: Massive

Composition: Diopside (clinopyroxene), plagioclase, almandine garnet, vesuvianite

Description: A metamorphic facies containing rocks that have undergone high temperature contact metamorphism

Rockname: Pyroxene Peridotite
Group: Igneous
Family: Peridotites
Texture: Crystalline
Structure: Massive
Composition: Olivene, augite (clinopyroxene), enstatite (orthopyroxene), hypersthene
Description: A type of ultramafic plutonic rock

Rockname: Pyroxenite
Group: Igneous
Family: Pyroxenites
Texture: Crystalline
Structure: Massive
Composition: Augite (clinopyroxene), enstatite (orthopyroxene), hypersthene, hornblende (amphibole)
Description: A type of ultramafic plutonic rock

Rockname: Quartz Alkali Feldspar Syenite
Group: Igneous
Family: Syenites
Texture: Crystalline
Structure: Massive
Composition: Orthoclase (K feldspar), quartz, biotite, hornblende (amphibole), pyroxene
Description: Dominant felsic mineral is K-feldspar. Lacks quartz

Rockname: Quartz Alkali Feldspar Trachyte
Group: Igneous
Family: Trachytes
Texture: Porphyritic

Structure: Massive, flow structure
Composition: Sanidine (K feldspar), quartz, biotite
Description: Lava flows and intrusives of felsic composition

Rockname: Quartz Anorthosite
Group: Igneous
Family: Anorthosites
Texture: Crystalline
Structure: Massive
Composition: Plagioclase, quartz, minor pyroxenes
Description: A felsic plutonic rock with plagioclase the major felsic mineral

Rockname: Quartz Arenite
Group: Sedimentary
Family: Sandstones
Texture: Clastic, grains 0.0625 to 2mm
Structure: Massive, bedded, laminated
Composition: Quartz, clay < 15%
Description: A quartzose sandstone with < 15% clay matrix

Rockname: Quartz Diorite
Group: Igneous
Family: Diorites
Texture: Crystalline
Structure: Massive
Composition: Plagioclase, quartz, hornblende (amphibole), pyroxenes, biotite, quartz
Description: A mafic plutonic rock

Rockname: Quartz Dolerite
Group: Igneous
Family: Dolerites
Texture: Crystalline, fine grained

Structure: Massive

Composition: Plagioclase, pyroxenes, quartz

Description: Usually occurs as dykes

Rockname: Quartz Gabbro

Group: Igneous

Family: Gabbros

Texture: Crystalline

Structure: Massive

Composition: Plagioclase, hornblende (amphibole), pyroxenes, biotite, quartz

Description: A mafic plutonic rock

Rockname: Quartz Latite

Group: Igneous

Family: Latites

Texture: Porphyritic

Structure: Massive, flow structure

Composition: Plagioclase, sanidine (K feldspar), quartz, biotite, hornblende (amphibole), pyroxenes

Description: Lava flows and intrusives of felsic composition

Rockname: Quartz Monzodiorite

Group: Igneous

Family: Granites

Texture: Crystalline

Structure: Massive

Composition: Plagioclase, orthoclase (K feldspar), quartz, biotite, hornblende (amphibole), pyroxenes

Description: A mafic plutonic rock

Rockname: Quartz Monzonite

Group: Igneous

Family: Syenites
Texture: Crystalline
Structure: Massive
Composition: Plagioclase, K- feldspar, quartz, biotite, hornblende (amphibole), pyroxenes
Description: Dominant felsic mineral is K-feldspar. Lacks quartz

Rockname: Quartz Norite
Group: Igneous
Family: Gabbros
Texture: Crystalline
Structure: Massive
Composition: Plagioclase, quartz, hypersthene (orthopyroxene)
Description: A mafic plutonic rock

Rockname: Quartz Rich Granitoid
Group: Igneous
Family: Granites
Texture: Crystalline
Structure: Massive
Composition: Quartz > 60%, orthoclase (K feldspar), plagioclase, biotite, hornblende (amphibole)
Description: A granitoid with greater than 60% quartz

Rockname: Quartz Syenite
Group: Igneous
Family: Granites
Texture: Crystalline
Structure: Massive
Composition: Orthoclase (K feldspar), plagioclase, quartz, biotite, hornblende (amphibole), pyroxene
Description: Dominant felsic mineral is K-feldspar. Lacks quartz

Rockname: Quartz Tholeiite

Group: Igneous

Family: Basalts

Texture: Porphyritic

Structure: Massive, flow structure

Composition: Plagioclase (laboradorite), pigeonite, quartz, hypersthene, augite (clinopyroxene), glass

Description: A type of Tholeiitic Basalt containing quartz.

Rockname: Quartz Trachyte

Group: Igneous

Family: Trachytes

Texture: Porphyritic

Structure: Massive, flow structure

Composition: Sanidine (K feldspar), plagioclase, quartz, biotite

Description: Lava flows and intrusives of felsic composition

Rockname: Quartz Wacke

Group: Sedimentary

Family: Sandstones

Texture: Clastic, grains 0.0625 to 2mm

Structure: Massive, bedded, laminated

Composition: Quartz, clay, clay 15% to 75%

Description: A quartzose sandstone with > 15% clay matrix

Rockname: Quartz, Feldspar, Biotite Schist

Group: Metamorphic

Family:

Texture: Crystalline

Structure: Schistose

Composition: Plagioclase, quartz, orthoclase (K feldspar), hornblende, biotite, muscovite,

silliminite
Description: Schistose variety of gneissic rocks

Rockname: Quartzite
Group: Metamorphic
Family:
Texture: Recrystallized
Structure: Massive, bedded
Composition: Quartz
Description: A general term used for rocks consisting entirely of quartz

Rockname: Quartzite Sedimentary
Group: Sedimentary
Family:
Texture: Clastic
Structure: Massive, bedded
Composition: Quartz
Description: A general term used for rocks consisting entirely of quartz

Rockname: Quartzolite
Group: Igneous
Family:
Texture: Crystalline
Structure: Massive
Composition: Quartz, mafic minerals
Description: Plutonic rocks with felsic component consisting of greater than 90% quartz.

Rockname: Rhyolite
Group: Igneous
Family: Rhyolites
Texture: Porphyritic
Structure: Massive, flow structure

Composition: Sanidine (K feldspar), plagioclase, quartz, biotite, , minor mafic minerals

Description: Lava flows and intrusives of felsic composition

Rockname: Rudstone

Group: Sedimentary

Family: Limestones

Texture: Clastic, grain supported fabric, >10% grains >2mm

Structure: Massive, bedded

Composition: Fossils, clasts, micrite, sparite, calcite, dolomite, quartz

Description: A very coarse, grain supported limestone

Rockname: Sandstone

Group: Sedimentary

Family: Sandstones

Texture: Clastic, grains 0.0625 to 2mm

Structure: Massive, bedded, laminated

Composition: Quartz, orthoclase (K feldspar), plagioclase, rock fragments, clay

Description: A general term for clastic rocks primarily consisting of quartz and feldspar grains and rock fragments cemented together

Rockname: Sanidinite Facies

Group: Metamorphic

Family:

Texture: Crystalline

Structure: Massive

Composition: Tridymite, crystabolite, sanidine, anorthoclase (K feldspar), plagioclase, pigeonite, mullite, glass, silliminite,

Description: A metamorphic facies containing rocks formed by very high temperature, low pressure pyrometamorphism during igneous intrusion

Rockname: Sannaite

Group: Igneous

Family: Lamprophyres
Texture: Porphyritic
Structure: Massive
Composition: Barkevite, kaersutite (amphibole), augite, olivene, biotite, orthoclase, foids, or > plag, felds > foids
Description: A mafic to ultramafic minor intrusive rock

Rockname: Schist
Group: Metamorphic
Family:
Texture: Crystalline, microcrystalline, recrystallized
Structure: Schistose
Composition: Biotite, muscovite, actinolite, chlorite, tremolite, serpentine
Description: A general term that describes schistose rocks of varying composition

Rockname: Scoria
Group: Volcanic
Family:
Texture: Porphyritic, highly vesicular
Structure: Massive
Composition: Plagioclase, augite (clinopyroxene), olivene, hornblende, orthopyroxene, glass
Description: A general term for highly vesicular volcanic rocks consisting mostly of gas bubbles

Rockname: Serpentinite
Group: Metamorphic
Family:
Texture: Crystalline, microcrystalline
Structure: Schistose, foliated, massive
Composition: Serpentine
Description: Formed by alteration of olivene rich ultramafic rocks and hydration of upper mantle peridotites

Rockname: Shale
Group: Sedimentary
Family: Mudstones
Texture: Clastic
Structure: Laminated
Composition: Clay, quartz, glauconite, pyrite
Description: A laminated mudstone deposited in lakes and seas

Rockname: Sherghottites
Group: Meteorites
Family: Stony meteorites
Texture: Crystalline
Structure: Massive
Composition: Pyroxenes, plagioclase
Description: Found after meteorite showers

Rockname: Shonkinite
Group: Igneous
Family:
Texture: Crystalline, coarse grained
Structure: Massive
Composition: Augite, K feldspar, foids (usually nepheline, minor olivene, biotite or hornblende
Description: A variety of foid syenite containing > 60% mafic minerals

Rockname: Shoshonite
Group: Igneous
Family: Basalts
Texture: Porphyritic
Structure: Massive, flow structure
Composition: Augite (clinopyroxene), plagioclase, orthoclase (K feldspar), olivene, glass

Description: A potassic basaltic rock of intermediate composition containing K feldspar. A variety of Basaltic Trachyandesite defined chemically with Na2O - 2.0 <= K2O

Rockname: Silcrete

Group: Sedimentary

Family:

Texture: Cryptocrystalline

Structure: Massive, inhomogenous

Composition: Quartz

Description: Deposited on and near the surface by precipitation from ground waters during weathering of siliceous rocks. Sometimes with colored convolutions

Rockname: Siltstone

Group: Sedimentary

Family: Siltsones

Texture: Clastic, grains 0.0039 to 0.0625mm

Structure: Massive, bedded, laminated

Composition: Quartz, orthoclase (K feldspar), plagioclase, clay

Description: Deposited in seas, lakes and rivers

Rockname: Skarn

Group: Metamorphic

Family:

Texture: Crystalline

Structure: Massive

Composition: Diopside, wollastonite, grossular garnet, zoisite, vesuvianite

Description: Formed during contact metamorphism of limestones

Rockname: Slate

Group: Metamorphic

Family:

Texture: Recrystallized

Structure: Foliated with slaty foliation

Composition: Clay, quartz, orthoclase (K feldspar), plagioclase, muscovite, chlorite, biotite

Description: Formed by low grade metamorphism of argillaceous sediments

Rockname: Spessartite

Group: Igneous

Family: Lamprophyres

Texture: Porphyritic

Structure: Massive

Composition: Hornblende, augite, olivene, plagioclase, serpentine (sec), chlorite (sec), plag > or

Description: A mafic to ultramafic minor intrusive rock

Rockname: Staurolite schist

Group: Metamorphic

Family:

Texture: Crystalline

Structure: Schistose

Composition: Staurolite, quartz, chlorite, garnet

Description: A schistose rock chiefly consisting of staurolite and formed at medium metamorphic grades

Rockname: Sub Anthracite

Group: Sedimentary

Family: Coals

Texture: Organic

Structure: Massive, bedded, laminated

Composition: Vitrinite, inertinite, liptonite, mineral matter

Description: A high grade (high carbon content) coal slightly lower in rank than anthracite

Rockname: Sub Bituminous Coal

Group: Sedimentary

Family: Coals

Texture: Organic
Structure: Massive, bedded, laminated
Composition: Vitrinite, inertinite, liptonite, mineral matter
Description: A medium grade coal slightly lower in rank than bituminous coal

Rockname: Subarkose
Group: Sedimentary
Family: Sandstones
Texture: Clastic, grains 0.0625 to 2mm
Structure: Massive, bedded, laminated
Composition: Quartz, minor orthoclase (K feldspar) plagioclase, clay < 15%
Description: A quartzose sandstone with minor feldspar and <15% clay matrix

Rockname: Sublitharenite
Group: Sedimentary
Family: Sandstones
Texture: Clastic, grains 0.0625 to 2mm
Structure: Massive, bedded, laminated
Composition: Quartz, minor rock fragments, clay < 15%, rock fragments < 25%
Description: A quartzose sandstone with minor lithoclasts and < 15% clay matrix

Rockname: Syderophyres
Group: Meteorites
Family: Stony iron meteorites
Texture: Crystalline
Structure: Massive
Composition: Fe-Ni alloys, bronzite, tridymite
Description: Found after meteorite showers

Rockname: Syenite
Group: Igneous
Family: Syenites

Texture: Crystalline

Structure: Massive

Composition: Orthoclase (K feldspar), plagioclase, biotite, hornblende (amphibole)

Description: Dominant felsic mineral is K-feldspar. Lacks quartz

Rockname: Tectonic Breccia

Group: Metamorphic

Family:

Texture: Fragmental

Structure: Massive, foliated

Composition: Rock fragments, mineral fragments

Description: Low to high pressure fragmentation of rocks along fault zones

Rockname: Tektite

Group: Tektites

Family:

Texture: Amorphous

Structure: Massive

Composition: Glass

Description: Small, brown to dark green glassy rocks formed at high temperatures during meteorite impacts or by ablation

Rockname: Tephrite

Group: Igneous

Family: Tephrites

Texture: Porphyritic

Structure: Massive, flow structure

Composition: Plagioclase, clinopyroxene, foids

Description: A volcanic rock of alkali composition occurring as lava flows and intrusions

Rockname: Tephritic Phonolite

Group: Igneous

Family: Phonolites
Texture: Porphyritic
Structure: Massive, flow structure
Composition: Sanidine (K feldspar), plagioclase, foids, biotite, hornblende (amphibole), pyroxenes
Description: A volcanic rock of alkali composition occurring as lava flows and intrusions

Rockname: Teschenite
Group: Igneous
Family:
Texture: Crystalline
Structure: Massive
Composition: Olivene, augite, plagioclase, analcime
Description: Synonomous with Analcime Gabbro

Rockname: Theralite
Group: Igneous
Family:
Texture: Crystalline
Structure: Massive
Composition: Augite, plagioclase, nepheline, olivene (variable)
Description: Synonomous with Nepheline Gabbro

Rockname: Tholeiite or Tholeiitic Basalt
Group: Igneous
Family: Basalts
Texture: Porphyritic
Structure: Massive, flow structure
Composition: Plagioclase (laboradorite), pigeonite, hypersthene, augite (clinopyroxene), glass
Description: An important type of Basalt defined chemically as containing normative hypersthene

Rockname: Tillite

Group: Sedimentary

Family: Conglomerates

Texture: Clastic

Structure: Massive, bedded

Composition: Quartz, rock fragments, clay

Description: Deposited by glaciers

Rockname: Trachyandesite

Group: Igneous

Family: Andesites

Texture: Porphyritic

Structure: Massive, flow structure

Composition: Plagioclase, sanidine (K feldspar), augite (clinopyroxene), hornblende, plus/minus glass

Description: Intermediate between trachyte and andesite with equal amounts of plagioclase and K feldspar.

Rockname: Trachybasalt

Group: Igneous

Family: Basalts

Texture: Porphyritic

Structure: Massive, flow structure

Composition: Plagioclase (labradorite), sanidine (K feldspar), augite (clinopyroxene), hornblende, olivene, plus/minus glass

Description: A basic volcanic rock with labradorite and K feldspar.

Rockname: Trachydacite

Group: Igneous

Family: Rhyolites

Texture: Porphyritic

Structure: Massive, flow structure
Composition: Quartz, K-feldspar, plagioclase, bronzite
Description: A variety of Rhyolite

Rockname: Trachyte
Group: Igneous
Family: Trachytes
Texture: Porphyritic
Structure: Massive, flow structure
Composition: Sanidine (K feldspar), plagioclase, biotite, minor mafic minerals
Description: A felsic volcanic rock with K feldspar the major felsic mineral

Rockname: Tremolite Schist
Group: Metamorphic
Family:
Texture: Crystalline, microcrystalline
Structure: Schistose
Composition: Tremolite, actinolite
Description: Formed by low grade regional metamorphism of mainly pelitic sediments

Rockname: Troctolite
Group: Igneous
Family: Gabbros
Texture: Crystalline
Structure: Massive
Composition: Olivene, plagioclase
Description: A mafic plutonic rock

Rockname: Trondhjemite
Group: Igneous
Family: Granites
Texture: Crystalline

Structure: Massive

Composition: Plagioclase, quartz, minor biotite

Description: A variety of Tonalite with minor mafic minerals. Synonomous with Plagioclase Granite

Rockname: Tuff

Group: Igneous

Family: Tuffs

Texture: Fragmental, consolidated, grain size < 2mm

Structure: Massive, bedded, laminated

Composition: Rock fragments, crystal fragments, glass fragments

Description: Ejecta accumulated by volcanic activity

Rockname: Uncompahgrite or Pyroxene Melilitolite

Group: Igneous

Family: Melilites

Texture: Crystalline

Structure: Massive

Composition: Melilite, augite (clinopyroxene)

Description:

Rockname: Ureilites

Group: Meteorites

Family: Stony meteorites

Texture: Crystalline

Structure: Massive

Composition: Olivene, pigeonite

Description: Found after meteorite showers

Rockname: Urtite

Group: Igneous

Family: Foidolites

Texture: Crystalline

Structure: Massive

Composition: Nepheline > 70%, aegerine augite, no feldspar

Description: A leucocratic variety of Nephelinolite (foidolite)

Rockname: Verite

Group: Igneous

Family: Lamproites

Texture: Inequigranular, porphyritic

Structure: Massive

Composition: Phlogopite, diopside, olivene, glassl

Description: A type of lamproite

Rockname: Vitric Tuff (consolidated) or Vitric Ash

Group: Igneous

Family: Tuffs

Texture: Fragmental, grain size < 2mm

Structure: Massive, bedded, laminated

Composition: Glass fragments predominant

Description: Ejecta accumulated by volcanic activity

Rockname: Vogesite

Group: Igneous

Family: Lamprophyres

Texture: Porphyritic

Structure: Massive

Composition: Hornblende, augite, orthoclase, olivene, serpentine (sec), chlorite (sec), or > plag

Description: A mafic to ultramafic minor intrusive rock

Rockname: Wackestone

Group: Sedimentary

Family: Limestones

Texture: Clastic, mud supported fabric

Structure: Massive, bedded

Composition: Fossils, clasts, micrite, sparite, calcite, dolomite, quartz, >10% grains

Description: A mud supported limestone

Rockname: Websterite

Group: Igneous

Family: Pyroxenites

Texture: Crystalline

Structure: Massive

Composition: Orthopyroxenes, augite (clinopyroxene)

Description: A type of ultramafic plutonic rock

Rockname: Wehrlite

Group: Igneous

Family: Peridotites

Texture: Crystalline

Structure: Massive

Composition: Olivene, augite (clinopyroxene)

Description: A type of ultramafic plutonic rock

Rockname: Wolgidite

Group: Igneous

Family: Lamproites

Texture: Inequigranular, porphyritic, madupitic

Structure: Massive

Composition: Richterite, leucite-diopside, phlogopite groundmass (madupitic)

Description: A type of lamproite

Rockname: Wyomingite

Group: Igneous

Family: Lamproites

Texture: Inequigranular, porphyritic

Structure: Massive

Composition: Phlogopite, leucite, diopside

Description: A type of lamproite

Rockname: Zeolite Facies

Group: Metamorphic

Family:

Texture: Recrystallized

Structure: Massive

Composition: Clay, zeolites, mixed layer clay

Description: A metamorphic facies containing very low grade metamorphic rocks that have undergone burial metamorphism.

Index of Rocknames

Rockname	Rockname	Rockname
Achondrites	Biolithite	Dolostone
Actinolite Schist	Bituminous Coal	Dunite
Agate	Blastomylonite	Dust Tuff
Agglomerate	Boghead Coal	Eclogite
Alaskite	Boninite	Eclogite Facies
Albite Epidote Hornfels Facies	Boundstone	
Algal Boundstone	Breccia	Enderbite
Alkali Basalt	Calcarenite	Enstatatite
Alkali Feldspar Charnockite	Calcilutite	Enstatite Chondrites
Alkali Feldspar Granite	Calcirudite	Epidote Amphibolite Facies
Alkali Feldspar Rhyolite	Calcite carbonatite	Essexite
Alkali Feldspar Syenite	Calcrete	Eucrites
Alkali Feldspar Trachyte	Camptonite	Evaporite
Alnoite	Cancarixite	Feldspathic Greywacke
Alvikite	Cannel Coal	Fenite
Amphibolite	Carbonaceous Chondrite	Fergusite
Amphibolite Facies	Carbonatite	Ferrocarbonatite
Analcimite	Cataclasite	Fitzroyite
Andesite	Cedricite	Floatstone
Angrites	Chalk	Foid Diorite
Anhydrite	Charno-Enderbite	Foid Gabbro
Anorthosite	Charnockite	Foid Monzodiorite
Anthracite	Chassignites	Foid Monzonite
Arenite	Chert	Foid Monzosyenite
Argillite	Chlorite schist	Foid Syenite
Arkose	Chondrites	Foidites
Arkosic Arenite	Clinopyroxenite	Foidolite
Ash	Coal	Framestone
Atexite	Comendite	Gabbro
Aubrites	Conglomerate	Gabbronorite
Australites	Coquina	Gaussbergite
Bafflestone	Cordierite Hornfels	Geode
Banded Iron Formation	Crystal Tuff (consolidated) or Crystal Ash	Glaucophane Schist
Basalt	Dacite	Glaucophane Schist Facies
Basaltic Trachyandesite	Diabase	Gneiss
Basanite	Diatomite	Gossan
Bauxite	Diogenites	Grainstone
Beforsite	Diorite	Granite
Benmoreite	Dolerite	Granodiorite
Bindstone	Dolomite Carbonatite	Granulite

Rockname	Rockname	Rockname
Granulite Facies	Lawsonite Schist	Nephelinite
Greenschist	Lherzolite	Nephelinolite
Greenschist Facies	Lignite or Brown Coal	Norite
Greenstone	Limburgite	Noseanite
Greywacke	Limestone	Obsidian
Gypsum Rock	Lipartite	Octahedrites
Halite Rock	Litharenite	Oil Shale
Harzburgite	Lithic Arkose	Olivene Basalt
Hauynite	Lithic Greywacke	Olivene Gabbronorite
Hawaiite	Lithic Tuff (consolidated) or Lithic Ash	Olivene Hornblende Pyroxenite
Hexahedrites	Madupite	Olivene Melilite
Hornblende Gabbro	Malignite	Olivene Orthopyroxenite
Hornblende Hornfels Facies	Mamillite	Olivene Pyroxene Hornblendite
Hornblende Peridotite	Mangerite	Olivene Pyroxenite
Hornblende Pyroxenites	Marble	Olivene Tholeiite
Hornblendite	Meimechite	Olivene Uncompahgrite or Olivene Pyroxene Melilitolite
Hornfels	Melilite	Olivene Websterite
Hortschiefer	Melilitolite	Opdalite
Howardites	Melteigite	Orendite
Hyalomylonite	Mesosiderites	Orthoconglomerate
Hypersthene Alkali Feldspar Syenite	Meta-Basalt	Orthopyroxenite
Hypersthene Syenite	Meta-Basite	Packstone
Ignimbrite	Meta-Pelite	Pallasites
Ijolite	Miaskite	Paraconglomerate or Diamictite
Iodranites	Mica Schist	Peat
Ironstone	Microgabbro	Pegmatite
Italite	Microgranite	Pelite
Jotunite	Migmatite	Phonolite
Jumillite	Minette, Igneous	Phonolitic Tephrite
Kersantite	Minette, Sedimentary	Phosphorite
Kimberlite	Missourite	Phyllite
Komatiite	Monchiquite	Picrite
Kugdite or Olivene Melilitolite	Monzodiorite	Picrobasalt
Lamproite	Monzonite	Plagioclasite
Lamprophyre	Mudstone	Plagiogranite (Plagioclase Granite)
Lapilli tephra	Mudstone, Carbonate	Polzenite
Lapilli Tuff	Mugearite	Porphyry
Larvikite	Mylonite	Prehnite Pumpellyite Facies
Laterite	Nakhlites	Psammite
Latite	Natrocarbonatite	Pseudotachylite

Rockname	Rockname
Pyroclastic Breccia or Agglomerate	Silcrete
Pyroxene Hornblende Gabbronorite	Siltstone
Pyroxene Hornblende Peridotite	Skarn
Pyroxene Hornblendite	Slate
Pyroxene Hornfels	Spessartite
Pyroxene Hornfels Facies	Staurolite schist
Pyroxene Peridotite	Sub Anthracite
Pyroxenite	Sub Bituminous Coal
Quartz Alkali Feldspar Syenite	Subarkose
Quartz Alkali Feldspar Trachyte	Sublitharenite
Quartz Anorthosite	Syderophyres
Quartz Arenite	Syenite
Quartz Diorite	Tectonic Breccia
Quartz Dolerite	Tektite
Quartz Gabbro	Tephrite
Quartz Latite	Tephritic Phonolite
Quartz Monzodiorite	Teschenite
Quartz Monzonite	Theralite
Quartz Norite	Tholeiite or Tholeiitic Basalt
Quartz Rich Granitoid	Tillite
Quartz Syenite	Trachyandesite
Quartz Tholeiite	Trachybasalt
Quartz Trachyte	Trachydacite
Quartz Wacke	Trachyte
Quartz, Feldspar, Biotite Schist	Tremolite Schist
Quartzite	Troctolite
Quartzite Sedimentary	Trondhjemite
Quartzolite	Tuff
Rhyolite	Uncompahgrite or Pyroxene Melilitolite
Rudstone	Ureilites
Sandstone	Urtite
Sanidinite Facies	Verite
Sannaite	Vitric Tuff (consolidated) or Vitric Ash
Schist	Vogesite
Scoria	Wackestone
Serpentinite	Websterite
Shale	Wehrlite
Sherghottites	Wolgidite
Shonkinite	Wyomingite
Shoshonite	Zeolite Facies

Minerals and Rocks
by Longwell, C.R., Knopf, A, Flint, R.F. (1934)

Introduction

Below a thin, ragged mantle of soil and superficial material, the Earth's outermost shell is made up of rocks. Most of these rocks are in turn made up of minerals. As the rocks are the chief documents in which the geologic history of the Earth is written, they be- come deeply interesting when regarded from this point of view. In order to penetrate their meaning and to understand them as historical records we must be able to recognize the minerals that make up the rocks. A mineral is a substance the product of inorganic nature, that is characterized by distinctive physical properties and a composition expressible by a chemical formula. Minerals are composed of chemical elements. A few consist of single elements, such as native gold and silver, as these metals are termed when they occur in elementary state in nature, or diamond and graphite, both of which are crystalline forms of the element carbon. Diamond and graphite illustrate in the most striking way possible what is meant by a mineral. Although both are identical in chemical composition, yet each is a distinct mineral because each has its own characteristic physical properties: diamond is transparent and is the hardest substance known, whereas graphite is opaque and is nearly the softest substance known. Most minerals, however, are made up of two or more chemical elements united in such a way that the product of the union differs greatly in its properties from those of the elements composing it.

Character of Minerals

Chemical Composition

A few minerals have an invariable chemical composition; but most of them have a variable composition which, however, can be expressed by a chemical formula. Quartz, one of the most abundant minerals, has a fixed composition, expressed by the chemical formula SiO_2 which is a sort of shorthand saying that one atom of silicon is united with two atoms of oxygen; in short, quartz, regardless of where obtained or how formed, is essentially 100 per cent (silica). Sphalerite from which most of the world's zinc is obtained, is a mineral of variable composition, which is indicated by writing its formula thus: Zn,FeS thereby indicates g that in this mineral an atom of iron can proxy for an atom of zinc, The various minerals react differently to chemical reagents, and these reactions are one of the means used in identifying minerals. It is beyond the scope of this book to explain how minerals are identified by their chemical behavior, but, many textbooks of mineralogy treat the subject fully.

Physical Characters

Nearly all minerals are crystalline that is to say, they are built up of atoms that are organized in definite geometric arrangements. A few minerals are amorphous (non-crystalline). Under favorable conditions of growth most minerals form crystals A crystal is a solid that is bounded by smooth plane surfaces called faces whose arrangement is governed by the internal structure of the mineral. The crystals of any particular mineral have forms that are more or less characteristic. For instance, the mineral pyrite frequently crystallizes in cubes. Garnet commonly occur as twelve-sided crystals known as dodecahedrons. The recognition of these crystal forms helps in identifying minerals.

Structure of Minerals. The structure of minerals generally refers to their outward shape and form. The following descriptive terms are used, some of which are self-explanatory: crystallized ,occurring as crystals or showing crystal faces; massive, not bounded by crystal faces: the antithesis of crystallized; columnar; fibrous (Fig. 6); botryoidal (Fig. 7), consisting of small rounded forms like closely bunched grapes; micaceous, occurring in thin sheets that can readily be split into thinner sheets; granular, in aggregates of coarse to fine grains; compact; earthy; oolitic, formed of small spheres resembling fish roe.

Cleavage and Fracture. The manner in which many minerals break or split is so characteristic that it helps greatly in identifying them. If they break so that smooth plane surfaces are produced, they are said to have a cleavage. Although this cleavage invariably occurs along planes, these planes are not necessarily parallel to the surface faces that bound the crystal.

Some minerals have but one cleavage; other have two, three, or even six different cleavage directions. The number of cleavage directions that a mineral has serves as an aid in determining the mineral. A fine example is the cubic cleavage of galena, which causes the galena to cleave in three planes at right angles to one another, so that it breaks up into small perfect cubes which can in turn be split into still smaller cubes, and so on (Fig. 8). Other examples are the rhombohedral cleavage of calcite, three planes not at right angles, so that the resulting cleavage fragments are rhombohedrons; and the cleavage of mica, in one direction only, the most remarkable cleavage in the whole mineral kingdom, by virtue of which the mica can be split into sheets of indefinite thinness. If a mineral has no cleavage, then the nature of its broken surface, its fracture, is more or less distinctive. The fracture of a mineral is conchoidal, if the surface of a fracture is curved like the interior of a clam shell; fibrous or splintery if it is like that of wood; uneven or irregular, if the surface is rough.

Color. The color of a mineral is one of its most conspicuous features. A few minerals have a distinctive color that serves as a ready mean s of identification. For example, the golden-yellow of chalcopyrite the lead-gray of galena, the black of magnetite, are striking properties of these minerals. The golden-yellow color of chalcopyrite, together with a test for soft brittle character, practically surfaces to identify chalcopyrite, but unhappily few minerals can be identified so easily. Surface alterations are likely to change the color of a mineral, as shown by the golden-yellow tarnish frequently seen on pyrite. To observe the true, intrinsic color of a mineral a fresh surface must be examined. Moreover, many minerals vary in color in the different specimens. This is due to a difference in composition such as an increased amount of iron in sphalerite, with the consequent darkening in color of the mineral: or to impurities such as the red color given to quartz by admixed hematite. Other minerals, such as fluorite (colorless, green, blue, violet), although having no perceptible variation in composition, show a wide range in color, the result of containing some foreign constituent in infinitesimal amount in a state of extremely minute subdivision evenly distributed through them.

A Typical Mineral Description

Name: Hornblende
Group: Amphibole
Formula: $(Na,K)0-1(Ca2(Mg,Fe2+,Fe3+,Al)5[Si6-7Al2-1)22](OH,$
Crystal System: Monoclinic
Color: Green, black, dark green, yellow-brown, brown
Opacity: Translucent
Luster: Vitreous
Streak: Colorless
SGLow: 3.02
SGHigh: 3.45
HardnessLow: 5
HardnessHigh: 6
Cleavage: 2
Direction: {110}, {100} good
Habit: Prismatic
Fracture: Uneven, brittle
Other: Most widespread of the amphiboles. Widespread within igneous and metamorphic rocks
Comments: A subgroup name in the amphibole group

Color of Powder or Streak. The color of the streak is an important aid in identifying minerals. The streak is a thin layer of the powder of the mineral obtained by rubbing the mineral on an unglazed porcelain place known as a streak plate. The color of the streak may be like that of the mineral, but surprisingly enough, the color of the streak of many minerals differs greatly from their body color. For example, some varieties of hematite are brilliantly black, but they give a red- brown streak, which positively identifies them as hematite.

Luster. The luster of a mineral is the appearance of its surfaces as determined by its inherent reflecting quality. Luster must not be confused with color, for two minerals of the same color can have to- tally different luster's, just as a black paint with a shiny finish, such as an enamel, differs in appearance from a black paint with a dull finish because it reflects light differently. The different kinds of luster are the following:

Metallic. Having the luster of a metal. Example: pyrite. Most minerals that give a dark or black streak have metallic luster.

Glassy Having the luster of glass. Example: quartz.

Resinous. Having the luster of yellow resin. Example: Sphalerite.

Pearly. Having the iridescence of pearl. Example: some varieties of feldspar.

Greasy. Looking as if covered with a thin layer of oil. Example: some varieties of massive quartz.

Silky. Like silk, as the result of a finely fibrous structure Example fibrous

Adamantine. Having a brilliant luster like that of a diamond.

Hardness of minerals. Minerals differ greatly in hardness, and the determination of this property is an important aid in identifying them. The reactive hardness of a mineral is determined by comparing it with the hardness of a series of minerals that has been chosen as a standard scale The scale consists of the following minerals, each mineral being harder than those that precede it in the scale.

Scale of Hardness
1. Talc
2. Gypsum
3. Calcite
4. Fluorite
5. Apatite
6. Orthoclase
7. Quartz
8. Topaz
9. Corundum
10. Diamond

The relative hardness of a mineral in terms of this scale is deter- mined by finding which of these minerals it can scratch and which it cannot scratch. In determining hardness the following precautions must be observed. A mineral that is softer than another may leave a mark on the harder one which can be mistaken for a scratch. The mark can be rubbed off. however, whereas a true scratch is permanent. Some minerals are commonly altered on the surface to material much softer than the original mineral The physical structure of a mineral may prevent the correct determination of its hardness. For instance, if a mineral is powdery, finely granular, or splintery in structure it can apparently be scratched by a mineral much softer than itself. It is always advisable when making the hardness test to confirm the test by reversing the procedure, that is, by rubbing the mineral of unknown hardness on the material of known hardness. The following materials serve as additions to the above scale. The finger nail is a little over 2 in hardness, as it can scratch gypsum but not calcite. A copper coin is slightly above 3 in hardness, as it can scratch calcite but not fluorite The steel of an ordinary pocket knife just exceeds 5, and ordinary glass has a hardness of 5.5.

Specific Gravity. The specific -gravity of a substance is expressed as a number that indicates how many times heavier a given volume of the substance is than an equal volume of water. Minerals range in specific gravity between 1.5 and 20.0. The great majority range between 2.0 and 4.0. There are various instruments by which the specific gravity of a mineral can be determined accurately, but ordinarily it is sufficient to judge the weight of a fair-sized piece in the hand. After some experience rather small differences in specific gravity can be detected in this way, and the specific gravity of a mineral can be roughly estimated.

Common Minerals

A few of the more common minerals are described on the following pages. The student should compare these descriptions with as many different specimens of the minerals as possible, and should note the form, color, and luster of each specimen and make the simple tests for hardness, streak, and specific gravity.

Magnetite. An oxide of iron, Fe_3O_4. Physical Characters. Black; metallic luster. Streak black. Hardness 6. Specific gravity 5.17. Strongly magnetic, hence its name. Granular or massive; fairly common in octahedral crystals (Fig. 10). Occurrence. Is a valuable iron-ore mineral, containing 72 per cent of iron. It is mined in the Adirondacks, New Jersey, Pennsylvania, and many other parts of the world. It is common as a minor constituent in

Pyrite. Iron sulphide, FeS_2 Physical Characters. Pale brass-yellow, but some specimens are tarnished to deeper shades of yellow. Streak black Hardness 6 to 6.5 unusually hard for a sulphide Specific gravity about 5. Generally granular Common as crystals, especially as cubes whose faces are marked with fine parallel lines, or striae (Fig. 4). Occurrence. The most common sulphide mineral. Occurs in many rocks and is an important vein mineral. May carry small amounts of gold or copper and so become an ore of both these metals. Is not used as an ore of iron, but as a source of sulfur in the manufacture of sulfuric acid. Its presence in building stones detracts from their value, as its oxidation produces not only iron-oxide stains but also sulfuric acid, which causes the stones to disintegrate.
rocks, particularly in the darker-colored igneous rocks. The black sand of the seashore is largely magnetite.-

Hematite. The ferric oxide of iron, Fe_2O_3. Physical Character Dark steel-gray to iron-black; brilliant metallic luster (except in earthy specimens). Streak light to dark red-brown (Indian red); color of streak distinguishes it from limonite. Hardness 5.5 to 6.5. Specific gravity about 5. Granular micaceous; earthy (in this form it is red). Rarely in crystals.
Occurrence: Hematite is widely distributed in rocks and is the most abundant ore mineral of iron; it contains 70 per cent of iron. More than nine-tenths of the iron produced in the United States comes from this mineral. The chief districts are near the shores of Lake Superior in Michigan, Wisconsin, and Minnesota. Other important districts are in northern Alabama and eastern Tennessee. Earthy hematite is the pigment that gives many sandstone's their red color. It is used also in red paints and as a polishing material.

Limonite. Hydrous ferric oxide, Fe_2O_3-H_2O. Physical Character Dark brown to nearly black Streak yellowish-brown, which distinguishes it from hematite. Hardness 5 to 5.5.. Specific gravity about 4. Common as masses that resemble compact bunched of grapes (botryoidal structure [Fig. l]); if broken open, these masses generally have a radiating fibrous structure; occurs also in stalactitic forms resembling icicles earthy. The term limonite is restricted to the amorphous and earthy forms, and the crystalline forms are called goethite. Occurrence. Limonite is a valuable source of iron, but contains less iron than magnetite and hematite. It is a common mineral formed by the alteration of previously existing minerals that contain iron.

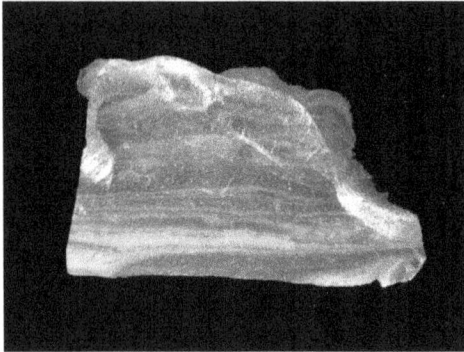

Jasper and Pyrite

Amethyst

Ordinary iron rust is limonite. It gives brown, orange, and yellow colors to many weathered rocks, to some non- weathered sedimentary strata, and to many soils.

Chalcopyrite (Copper Pyrite). Copper-iron sulphide, $CuFeS_2$. Physical Characters Golden-yellow; generally tarnished to bronze or iridescent colors. Streak greenish-black. Hardness 3.5, hence much softer than pyrite. Specific gravity 42. As a rule massive, rarely in crystals Occurrence. An abundant and valuable ore-mineral of copper, containing 34 per event of the metal. Occurs intimately distributed in vein deposits with many other sulphide minerals.

Sphalerite. Zinc sulphide, ZnS, when ideally pure; generally contains some iron, as indicated by the formula $ZnFeS$. Physical Character Commonly yellow-brown to dark-brown, being darker in the varieties containing more iron. Resinous to submetallic luster Hardness 3.5 to 4. Specific gravity about 4. White to yellow and brown streak, of lighter shade than the mineral itself. Has brilliantly Bashing cleavage planes sloping in six different directions. As a rule massive. Occurrence. The most important source of zinc. Widely distributed, but generally in veins or irregular bodies in limestone. Associated generally with galena, pyrite, and chalcopyrite.

Galena Lead sulphide PbS. Physical Characters. Lead-gray. Bright metallic luster. Streak grayish-black Hardness 2.5 (soft). Specific gravity about 7.5 (" very heavy "). Perfect cleavage in three planes at right angles to each other, forming cubes Fig. 8 (not visible, however, in finely granular specimens). Occurs in natural cubic crystals, but massive and granular aggregates are more common. Occurrence. Is the chief source of lead; contains 87 per cent of the metal. Some contains silver and serves as an ore of that metal. Commonly occurs with zinc minerals.

Calcite. Calcium carbonate, $CaCO_3$. Physical Characters. Generally white or colorless. Also variously tinted gray, red, green, and blue. Commonly opaque or translucent; rarely transparent Hardness 3. Specific gravity 2.7. Perfect cleavage in three planes at oblique angles to each other (rhombohedral cleavage [Fig. 9]), giving rhombic-shaped faces. well-formed crystals are common. Effervesces freely on application of cold acid, because of the copious liberation of the gas carbon dioxide. This test serves to distinguish calcite from dolomite, another common carbonate, which does not effervesce under these conditions. Occurrence. A very common mineral. Is the chief constituent of limestone's and marbles; also common in veins. Used in the manufacture of lime, plasters, and cement as a metallurgic flux, and in chemical industries.

Dolomite. Carbonate of calcium and magnesium, $CaMg(CO_3)_2$. Physical Characters. Generally white or gray; rarely flesh-colored. Opaque to translucent. Hardness 8.5 to 4 (harder than calcite). Perfect cleavage in three planes not at right angles to each other (rhombohedral cleavage). Specific gravity 2.9. Glassy to pearly luster. Does not effervesce on application of a drop of cold acid unless the acid is placed on a scratched or powdered surface. In this respect it differs from calcite. Occurrence. Composes the common rock known as dolomite; also dolomite marble. Occurs also as a vein mineral. In the rock form, it is used as a building and ornamental stone, for the manufacture of some cements, as s source of magnesia for refractory substances and as agricultural lime.

Gypsum. Hydrous calcium sulfate, $CaSO_4 \ 2H_2O$. Physical Characters. Usually white or colorless. Hardness 2 easily scratched with the finger nail). Specific gravity about 2.8. Has one perfect cleavage; an- other imperfect cleavage is visible in some specimens. Occurrence. Is widely distributed in sedimentary rocks. In place forma thick beds, commonly interstratified with limestone and shale. Generally occurs in association with salt beds. Is chiefly used for the production of plaster of Paris.

Halite (Common Salt). Sodium chloride, $NaCl$ Physical Characters. White or colorless. Hardness 2.5. Specific gravity 2.1. Perfect cleavage in three planes at right angles to one another (cubic cleavage). Transparent to translucent. Salty taste. Generally in cubic crystals or in masses showing cubic cleavage. Occurrence. In thick beds interstratified with sedimentary rocks and associated with gypsum. Used for cooking and preservative purposes; also extensively in chemical industry.

Quartz. Silicon dioxide, SiO2. Physical Characters. Colorless or white; but many varieties are colored by impurities, yellow, red, pink, amethyst, green, blue, brown, black. Glassy luster. Transparent to opaque. Hardness 7. Specific gravity 2.65. In contrast to most common minerals, quartz shows no cleavage; has conchoidal fracture. Commonly in hexagonal crystals similar to Fig. 11. The triangular faces at the ends of the crystals are usually smooth, whereas the rectangular faces between the ends are horizontally striated. Also massive. Varieties. There are many varieties of quartz to which different names are given. A few are as follows: rock crystal which is colorless quartz, commonly in distinct crystals; amethyst quartz colored purple or violet; rose quartz, usually massive with a pink color; smoky quartz, quartz of a smoky yellow, brown, or almost black color; chalcedony, finely fibrous variety, translucent with a waxy luster; agate, a variegated chalcedony delicately banded in different colors; jasper, extremely fine-grained quartz colored red by admixed hematite. Occurrence. Quartz is one of the most common minerals. It is most abundant as a constituent of granite, in which it resembles bits of window glass. It is also the most common vein mineral. It makes up the largest part of most sands and sandstone's.

Garnet. There are several garnets, which differ from one another in the elements they contain. They are all silicates with analogous chemical formulas. The most common garnet is a red variety (almandine), containing ferrous iron and aluminum, FesAl2(Si04) g. Other garnets contain magnesium, calcium, manganese, and ferric iron. Physical Characters. Color depends somewhat on the composition, but is an unsafe criterion; most commonly red (almandine) or brown. Also yellow, green, and black. Transparent to almost opaque. Hardness 7. Specific gravity 32 to 43 (varies with the chemical composition Generally well crystallized, either in a form showing 12 rhombic-shaped faces (dodecahedron, Fig. 5) or 24 trapezium- shaped faces (trapezohedron). Occurrence. Garnet is a widely distributed mineral, occurring most commonly in metamorphic rocks. Used as a semi-precious gem stone and, because of its hardness, as an abrasive material l

Orthoclase (Potassium Feldspar). Potassium-aluminum silicate, KAlSi2O4. Physical Characters. Colorless, white, gray, pink, and red; rarely green. Streak white, in spite of the diversity of colors. Hardness 6. Specific gravity 2.56. Has two good cleavages that make right angles with each other (whence the name of the mineral). Occurrence. The most common silicate. Widely distributed as a prominent rock constituent, occurring in rocks of many kinds, but most abundantly in granite and allied rocks. Also in large crystals and cleavage masses in what are known as pegmatite dikes. From these dikes it is quarried in large amounts for use in the manufacture of porcelain.

Plagioclase Feldspars. Sodium-calcium-aluminum silicates. Physical Characters. Various shades of gray, less commonly white. Tans- parent to opaque. Hardness 6. Specific gravity 2.6 to 2.76. Have two cleavages making nearly a right angle with each other, one of them (the basal cleavage being better than the other. Some specimens are distinguishable from orthoclase by having on their basal cleavage planes a series of stria (fine parallel lines, which resemble rulings made by a fine diamond point). Some cleavage surfaces, especially of the dark-gray variety, give a beautiful play of colors when specimen is rotated in good light. The white variety commonly occurs in thin-bladed crystals with curved surfaces and a pearly luster. Occurrence. In much the same manner as orthoclase. The plagioclase in gabbros is likely to be dark colored or black and therefore not easily distinguishable from the associated pyroxene.

Muscovite (White Mica). A complex silicate containing potassium and aluminum. Physical Characters Has a perfect cleavage in one direction, which allows the mineral to be split into exceedingly thin sheets or flakes. These sheets are flexible and elastic, ie., on release of pressure a bent sheet- has the power of resuming its original shape. Transparent and almost colorless in thin sheets. In thicker blocks, opaque with light shades of brown and green. Hardness 2 to 2.5. Specific gravity 2.76 to 3. Occurrence. A common rock-making mineral. It occurs in granite together with quartz and feldspar, and with the same minerals in pegmatite dikes It is characteristic of a series of rocks made up of abundant mica, in which it is arranged in parallel orientation, with the result that the rocks split in flakes and slabs parallel to the cleavage of the mica. These rocks are known as mica schists. Is used chiefly as an insulating materials l in the manufacture of electrical apparatus. There are many minor uses.

Biotite (Black Mica). A complex silicate containing potassium, magnesium, iron, and aluminum. Physical Characters. Perfect micaceous cleavage. Cleavage sheets and flakes are flexible and elastic. Generally dark-green, brown, or black. Thin sheets generally have a smoky color differing from the almost colorless muscovite). Hardness 2.5 to 3. Specific gravity 3. Occurrence. An abundant rock-making mineral, common in granites and many gneisses and schists.

Chlorite. A complex silicate containing magnesium and aluminum. Chlorite is a name given to a group of minerals that are prevailingly of green color and of broadly similar characters. Physical Characters. Perfect micaceous cleavage. Flakes are flexible but not elastic differing in this lack of elasticity from the micas). Green of various shades. Hardness 2 to 2.5. Specific gravity 2.65 to 2.96. Occurrence. A common rock-making mineral. The green color of many rocks is due to the presence of this mineral. This is particularly true of many schists and slates (green roofing slates).

Serpentine. Magnesium silicate , H4Mg3Si2O3. Physical Characters. Green of various shades: olive-green, yellow-green. ranging to blackish-green. Luster greasy or wax-like; silky when fibrous. Hardness 2.5 to 5, generally 4. Specific gravity 2.5 to 2.65. Usually massive but also fibrous or felted. The name serpentine is given also to the common rock com- posed largely of the mineral serpentine. Occurrence. A common mineral, widely distributed. Invariably an alteration product of some magnesia n silicate, chiefly olivine. It is the chief constituent of the rock called serpentine, some varieties of which are used as ornamental stone. The fibrous variety known as chrysotile is the principal source of asbestos.

Olivine. Silicate of magnesium and iron, (Mg,Fe) 2Si04. Physical Characters. Olive-green to yellowish-green; rarely brownish. Trans- parent to opaque. Hardness 6.5 to 7. Specific gravity 327 to 3.37. Glassy luster Conchoidal fracture, causing it to resemble yellowish-green quarts. Occurrence. Common as irregular grains in the dark, heavy granular igneous rocks gabbros and peridotites, and as distinct crystals in many basalts.

Pyroxene and Amphibole

These two abundant rock making minerals are similar in some respects, and their identification is based on crystal form. Consequently, it is difficult to discriminate them in most rocks, because good crystal forms are rare in rocks. However, it is well to study them separately under favorable conditions, in order to appreciate their differences as well as their points of similarity.

Pyroxene. A silicate containing chiefly calcium and magnesium; also varying amounts of aluminum, iron, and sodium. Physical Characters. Light- to dark-green to black, varying with the amount of iron; white in iron-free variety. Commonly opaque. Hardness 5 to 6. Specific gravity 3.1 to 3.6. In prismatic crystals with eight sides (Fig. 12a); in reality square prisms whose corners are truncated. The angle between alternate faces is therefore nearly 90'. These faces will fit into the corner of a box or tray. By means of these angles pyroxene can best be told from amphibole. Some specimens show a fair cleavage parallel to the faces lettered m in the figure, the angle between the cleavage faces being also nearly 90'. This cleavage is not visible in all specimens used for demonstration purposes. Occurrence. Pyroxene is a highly abundant rock-making mineral, occurring chiefly in the dark-colored igneous rocks. Rare in rocks that contain much quartz.

Amphibole. Silicate of calcium and magnesium with varying amounts of aluminum, iron, and sodium. Chemical composition is much like that of pyroxene. Physical Characters. Light- to dark-green to black, varying with amount of iron. Commonly opaque, Hardness 5 to 6. Specific gravity 2.93 to 3.8. Commonly in prismatic crystals with six sides. Figure 12b shows that the angles between the faces lettered m are 124 and 56' (very different from the corresponding angles in pyroxene). Has a good cleavage parallel to the faces lettered m. The difference in the form of the crystals and in cleavage angles and the fact that amphibole has the better cleavage are the chief outward distinctions between amphibole and pyroxene. Amphibole as a rule has a higher luster and yields smoother more continuous cleavage surfaces than does pyroxene. Some varieties of amphibole have long, needle-like crystals resulting in a fibrous structure. Pyroxene does not occur in this form. It should be remembered that the 56' and 124' with each other. The presence of well-formed crystal faces or cleavage surfaces is essential in order to distinguish between pyroxene and amphibole in hand specimens. Occurrence. Amphibole is an abundant rock-making mineral, occurring in biotite rich igneous and metamorphic rocks. Hornblende is a common dark variety of amphibole. Pyroxene and amphibole together with biotite are the dark minerals commonly occurring in many rocks. The first two can be distinguished from biotite by the fact that they occur in prismatic crystals that cannot be divided into thin elastic Hakes; that is, they lack the perfect cleavage of the micas. If present as small grains in a rock, they lack the high luster characteristic of flakes of biotite. They can be distinguished from chlorite by their much greater hardness as well as by their form and lack of micaceous cleavage.

Rocks

Most rocks are aggregates of minerals. Therefore they differ greatly in appearance and other properties, depending on what minerals are present, the number of minerals present and their relative abundance, the size of the mineral grains, and the way in which the minerals are associated. The kinds of rocks are many and the varieties endless, but, if classified according to their modes of origin, they fall into three major classes:

I.   Igneous rocks formed by the solidification of molten rock-matter, as exemplified by the rocks formed on the cooling of lava discharged from a volcano.

II.   Sedimentary rocks, most of which were formed by the settling of their substance as sediment from bodies of water such as streams lakes, and the sea.

III.   Metamorphic rocks, formed from pre-existing rocks by the development of new characters as the results of pressure, heat, or other geologic agents acting on them within the Earth's crust.

Characters Used in Identifying Rocks

The properties most useful in identifying rocks are structure, texture, hardness, and fracture. Color is sometimes useful, but is often misleading. Structure is a term used in describing the larger features of rocks. A layered or laminated structure generally indicates sedimentary origin. If a rock contains spherical or almond-shaped cavities or vesicles (" blowholes" formed by the liberated and expanding gases that were dissolved in molten rock-matter), it has s vesicular structure and is of igneous origin. If the vesicles become filled with minerals, the resulting structure is termed amygdaloidal (Fig. 18). Texture is the appearance of a rock as determined by the size, shape, and arrangement of the mineral grains of which it is built. The size of the grains determines the grain-size of the rock. If its grains are as large as peas, a rock is said to be coarse grained, or coarsely granular; if as small as the grains in granulated sugar, the rock is termed fine grained; and if so small that they cannot be distinguished as separate entities, the rock will seem to be a homogeneous substance and is said to be aphanitic. The shape and arrangement of the mineral grains with respect to one another produce s distinct pattern the fabric of a rock. If, for example, the grains are roughly of one size, the rock has an equigranular fabric; but if the grains are very unequal in size, the fabric is termed inequigranular. There are many fabrics, some of which are characteristic of the rocks in which they occur. In as much as texture (" appearance of the rock ") is the conjoint effect of grain-size and fabric, texture is customarily (and loosely) used for grain-size, for fabric, or for their conjoint effect. Certain textures help in the megascopic identification of rocks, i.e., without the aid of the microscope, which is all that is attempted here. The texture of a granite, so distinctive that it is termed the granitic texture, proves not only that the rock is of igneous origin, but also that it was formed under conditions of slow undisturbed cooling. A glassy texture proves that a rock having this texture is also of igneous origin, but that, unlike granite, it was formed by molten rock-matter suddenly solidifying, for glasses are the results of extremely rapid cooling. The clastic texture (clastic, from the Greek meaning broken) occurs in rocks that are made up of angular or more or less rounded fragments of minerals and rocks, and is characteristic of many sedimentary rocks. Other textures are described in connection with particular rocks. Hardness aids in identifying some rocks. any rocks resemble limestone, but the test for hardness with the knife-point serves at once to distinguish a limestone, the hardness of which is 3, from the much harder rocks that resemble it. Fracture is a less useful property. However, perfect conchoidal fracture characterizes the volcanic glasses (Fig. 14); and a semi- conchoidal fracture yielding shell-like fragments characterizes shales. The tendency of most metamorphic rocks to split into slabs or thin flakes is a valuable aid to their identification.

A Typical Rock Description

Name: Limestone
Rock Group: Sedimentary
Family: Limestones
Texture: Clastic, biogenic, chemical, pelletal, oolitic
Structure: Massive, bedded, laminated
Minerals: Fossils, clasts, micrite, sparite, calcite, dolomite, quartz
Occurrence: Basin

Other: A general term for carbonate rocks formed by chemical precipitation, mechanical accumulation or biological formation of calcium carbonate
Location: Lilydale, Vic
Formation: Lilydale Limestone
Age: Devonian

Igneous Rocks

As their name implies, igneous rocks were formed at high Temperatures, and they are defined as those rocks made by the cooling of molten matter that originated within the Earth. This molten matter when rising from the depths is more or less highly charged with gases; and these gases begin to escape as soon as the pressure on the liquid is reduced, and they are entirely eliminated when the liquid solidifies. The liquid rock-matter plus its content of dissolved gas is called magma. There are many kinds of magma at least forty and they have a wide range in composition.

Extrusive and Intrusive Rocks. Magma extruded at the Earth's surface solidifies on cooling to form extrusive rocks. Vastly greater quantities of magma than ever reached the surface have remained within the crust, however, and have solidified there under a cover consisting of the rocks of the upper part of the Earth's crust. This magma moved upward from the place where it originated : it is an intruder in the place it now occupies, hence the resulting rocks are termed intrusive Manifestly, such bodies of intrusive rock have become accessible to view only after they have been uncovered by erosion. Intrusive rocks were formed in a geologic environment that differs greatly from that in which extrusive rocks have formed. In an intrusive mass the magma cools under an insulating cover of rocks; hence its dissolved gases tend to be held until a late stage of solidification and it loses heat slowly, and therefore it solidifies slowly. In extrusive bodies, such as a lava flow the magma, becomes drastically chilled by exposure to the atmosphere even more so, by flowing into water and solidifies rapidly. As a result, most intrusive rocks differ greatly in appearance from extrusive rocks.

Here are some rock classification terms and their definitions:

RockGroup

Igneous – Crystalline rocks that originate from the Earth's interior and crystallize from molten magma.

Introduction to Geology

Metamorphic – Deformed and/or recrystallized rocks that have changed in composition and/or texture due to temperature and pressure changes in the Earth.

Sedimentary – Rocks composed of grains derived from pre-existing rocks and accumulated by physical processes, such as ocean and river basins, or chemically precipitated from water.

Meteorites – Rocks originating from space

Duricrust – Rocks formed on and near the surface due to weathering.

Structure

Massive - No structure

Foliated - Planar alignment of platy minerals

Schistose - Discontinuous planar alignment of minerals

Flow - Changing alignment of minerals in a flow pattern

Banded - With bands of different composition or color

Bedded/Laminated - Divided into separate layers

Texture

Crystalline - All crystalline textures, with a grain size > 0.2mm (eg. Granite, Gneiss)

Microcrystalline - Crystalline textures with a grain size < 0.2mm and > 0.01mm (eg. Hornfels)

Cryptocrystalline - Crystalline textures with a grain size < 0.01mm (eg. Agate)

Amorphous - Non crystalline, glassy (eg. Obsidian)

Porphyritic - Large crystals (phenocrysts) in a fine grained ground mass (eg. Basalt)

Recrystallized - Textures produced by partial recrystallization (eg. Meta-Basalt)

Inequigranular - Fine to macro - sized grains (eg. Kimberlite)

Fragmental - Composed of mineral and/or rock fragments (eg. pyroclastics)

Biogenic- Textures produced by organisms (eg. Limestone - Boundstone)

Organic-Textures produced by organic material (eg. coal)

Introduction to Geology

Clastic - Textures produced by mechanically accumulated grains cemented together (eg. Sandstone)

Chemical - Textures produced by chemical precipitation (eg. Anhydrite)

Occurrence

Plutonic - A general term for any large scale intrusive rocks

Volcanic - Extrusive and associated intrusive rocks

Regional Metamorphic - Metamorphic rocks occurring over large areas

Contact Metamorphic - Adjacent to intrusions

Fault/Shear Zone - Planar zones of brittle and/or ductile deformation

Basin - Sedimentary rocks in a sedimentary basin.

Texture and Composition of Igneous Rocks

The most obvious thing about an igneous rock, except perhaps its color, is its texture. By texture, as already explained, is meant the appearance of a rock as determined by the size, shape, and arrangement of its constituents. Most igneous rocks are made up of mineral grains, but some consist of glass, and some of glass and mineral grains. The grain-size is coarser the more slowly the magma cooled When the magma is extremely hot, the minerals dissolved in it cannot crystallize out; that is, the atoms and atomic groups in the magma can not arrange themselves to form organized solid compounds (the minerals). After the temperature has fallen far enough the minerals begin to separate from the magma, and, if the cooling is slow, they have time to grow to large size, thus forming a coarse-grained rock. But if cooling is rapid, more and more nuclei centers of crystallization form spontaneously, and, instead of a few such nuclei growing to large crystals, many begin to grow simultaneously; therefore none of them can attain large size, and consequently the resulting rock is fine grained. If cooling is still more rapid, the crystals remain so minute that they are not visible to the unaided eye, and the resulting rock is aphanitic Under conditions of extremely rapid cooling the magma solidifies into a homogeneous substance before any crystallization can occur. In this event the product is a glass sometimes called a natural glass. Porphyry: Porphyritic Texture. So far all the mineral grains in a given rock have been tacitly assumed to be of uniform size, i.e., That the rock is equigranular . Not all igneous rocks, however, are equigranular. Many of them consist of grains of two markedly contrasting sizes: in part of large conspicuous crystals and in part of much smaller grains, which form a matrix inclosing the large crystals. An igneous rock of this texture is said to be porphyritic. The matrix is called the ground- mass, and the large crystals imbedded in the groundmass are the phenocrysts, the easily discernible crystals. Porphyritic rocks are abundant and of many kinds. Some have medium- grained groundmasses; others have fine-grained, aphanitic, or glassy ground- masses as exemplified by the lavas, which as a rule are porphyritic rocks having aphanitic groundmasses. The phenocrysts also range greatly in size from those several in the s in diameter to those so small as to be barely visible to the unaided eye, and they range from sparse to abundant. In all porphyritic rocks, however, the phenocrysts contrast conspicuously in size with the grains that make up the groundmass; and this contrast in size is the essential feature of a porphyry. The porphyritic texture is not a contrast of colors; thus a rock made of grains of light-colored quartz and feldspar and containing a few crystals of black mica in marked color contrast, the grains of all three minerals being of about the same size, is not a porphyry.

Factors Determining Grain-Size. Reduced to its fundamentals, grain-size is determined (1) by the viscosity of the magma during the time the minerals are growing and (2) by the time available for them to grow. Low viscosity increases the mobility of the atoms while they are getting together to build up the growing crystals, and adequate time is necessary for them to travel to the growing crystals. Viscosity is enormously affected by the mica composition of the magma. Silicic magma (those high in silica) are several hundred thousand times as viscous as are the subsilicic magmas (those low in silica) . The effect of this great difference in viscosity is impressively illustrated by the strongly contrasting behavior of these magmas when erupted at the Earth's surface. Subsilicic magmas require such drastic chilling to form glasses that subsilicic glasses are rare, whereas silicic magmas, because of their great viscosity and consequently feeble tendency to crystallize, commonly produce silicic glasses. No subsilicic counterpart of the large, thick flow of silicic glass (obsidian) at Obsidian Cliff, Yellowstone National Park, occurs anywhere in the world. The presence of gases especially the water contained in magmas , decreases the viscosity of the magma and thereby promotes a coarser crystallization to an astonishing degree. An intrusive body of mag- ma, cooling and solidifying under a cover of rocks, is likely to retain its gases to a late stage and consequently to develop a coarsely crystal- line texture. Depth in the crust and especially gas-tightness of the surrounding rocks are therefore important factors in determining the texture of igneous rocks. A silicic magma increases enormously in viscosity when its gas content escapes; hence the very great contrast in appearance between the rocks formed in a gas-tight environment within the crust and the rocks formed from the same kind of magma erupted at the Earth's surface, where of course the gas content escapes freely. However, if the subterranean magma solidifies surrounded by confining rocks that are fissured or otherwise allow the gases to escape, it takes on a texture like that of surface rocks. For example, in the Rocky Mountains of Colorado, certain intrusive masses are coarse grained, whereas others formed at the same time, but deeper in the crust, are porphyritic; evidently depth was not the controlling factor here in developing coarsely granular texture. The importance of depth in determining a coarse-grained texture in igneous rocks has in the past been greatly exaggerated.

Minerals Common in Igneous Rocks. The minerals that make up most of the igneous rocks are the following:

Light-colored Group
(Feldspar Group)
Orthoclase feldspar Plagioclase feldspars
Quartz

Dark-colored Group
(Ferromagnesian Group)
Biotite (black mica)
Pyroxene
Hornblende
Olivine
Magnetite

The rocks in which the light-colored minerals predominate are light in color and light in weight, i.e., they are of low specific gravity. The rocks predominantly composed of dark ferromagnesian minerals are dark in color and heavy in weight. The range in specific gravity from 2.67 for the average granite to 3.0 for gabbro is not large, but surfaces after some experience to serve as an aid in identification. Although each rock has been placed in the table in a separate compartment, in nature no rock variety is as sharply delimited from its neighbors as it seems to be in the table. For example, there are transitional varieties between granite and diorite and between granite and granite porphyry. In more detailed classifications these intermediate rocks receive special names. No hard and fast boundaries set off any of the so-called rock species. These facts often make it difficult to classify a given rock. Great difficulties are presented by the finer- grained rocks. When the accurate identification of a rock becomes s matter of high importance, recourse must be had to expert microscopic examination.

Equigranular Rocks

The equigranular rocks are the results of slow cooling, combined with retention of the gas content in the magma until it has nearly solidified. Typically they occur in intrusive bodies, especially in those of great size. In such voluminous masses, cooling was necessarily slow, and the pressure was sufficient to keep the gases within the magma and to allow them to exercise their power of promoting coarse crystallization.

Granite. As can be seen from the scheme of classification, granite consists largely of quartz and feldspar (mainly of the variety orthoclase) . It contains also as a rule some black mica (biotite); less commonly it contains hornblende, or hornblende and biotite together. All these component minerals are roughly of the same size, and hence granite is said to be equigranular. The minerals began to separate from the magma in a definite order: first the dark minerals, hornblende and biotite; then the feldspars; and last the quarts. The dark minerals, being the first to crystallize, were not hampered in their growth by the presence of any neighbors, and so are generally in the form of sharply defined crystals; and the feldspars, having begun to grow later are less well crystallized, for where they abutted upon the earlier-formed dark minerals their freedom to grow was hampered. As the quartz was the last mineral to separate from the magma, it had to take what space was left, and it is therefore molded around the earlier minerals and occupies the angular interspaces between them, This habit of the quartz produces an intimately interpenetrating and interlocking arrangement.

Diorite. Diorite is an equigranular igneous rock composed of feldspar and one or more dark minerals, in which the feldspar is more abundant than the dark minerals. The feldspar is mainly plagioclase, but unless the characteristic striae on the cleavage planes can be seen it is generally difficult to recognize the plagioclase with the unaided eye. The dark minerals are biotite, hornblende, or pyroxene, occurring either singly or together. Syenite is much like granite in composition, differing only in containing little or no quartz; hence it is classed here with diorite. It is not common, nor does it as a rule occur in large masses compared with the enormous bodies of granite.

Gabbro. Gabbro differs from diorite in that the feldspar is sub- ordinate and the dark minerals predominate. Hornblende, pyroxene, and olivine are the common dark minerals, occurring singly or together; biotite, though present in some gabbros, is distinctly uncommon. Because the dark minerals predominate, gabbros are dark and of high specific gravity. Dolerite is a convenient term. for the basic rocks that in grain-size are intermediate between basalts and gabbros.

Peridotite. Peridotite is composed wholly of ferromagnesian minerals, with olivine predominating.

Pyroxenite, as its name implies, is composed wholly of pyroxene, and hornblendite consists entirely of hornblende As a rule hornblendite and pyroxenite form bodies of small size; nevertheless, in places, as at the remarkable platinum deposits recently discovered in South Africa, pyroxenite occurs in vast volume.

Porphyritic Granular Rocks

Rocks of this class are distinguished by the fact that they contain phenocrysts imbedded in a groundmass so coarse grained that its component minerals can be recognized by the unaided eye. The phenocrysts in most of these porphyries are abundant making up half or more of the bulk of the rock. If the volume of the phenocrysts exceeds 75 per cent, the porphyry becomes to the unaided eye indistinguishable from the corresponding granular rock.

Granite Porphyry; Diorite Porphyry etc. Typical granite porphyry contains conspicuous crystals of feldspar, quartz, and biotite set in a granitic groundmass. As its name implies, its composition is like that of granite it differs from granite in having phenocrysts and a groundmass whose grain-size on the average is finer than the grain- size of the average granite. Diorite porphyry differs from granite porphyry in the absence of quartz phenocrysts and the prevalence of phenocrysts of plagioclase feldspar.

Porphyritic Aphanitic Rocks

The rocks of this class are generally of volcanic origin. Extruded upon the Earth's surface, the magmas from which they were formed have cooled rapidly. They are characterized by the occurrence of porphyritic crystals (Fig. 16) set in a groundmass that is either so fine grained as to be irresolvable by the unaided eye or else is partly or wholly glassy.

Rhyolite. Rhyolite contains phenocrysts of feldspar, quartz, and biotite, and rarely of hornblende, set in an aphanitic groundmass. These phenocrysts range in number within the widest limits, so that there is a complete transition from non porphyritic to highly porphyritic rhyolite. The colors typically are light, ranging from white to gray, pink, red, and purple. Rhyolites and andesites that have inconspicuous phenocrysts or few or no phenocrysts are termed felsites.

Andesite. Andesites are of many colors, but in general they are darker than the rhyolites; dark gray is common. They are transitional on the one hand into rhyolites; on the other, into basalt. The average or typical andesite occupies the intermediate position. The darker andesites are like basalts in appearance, but unlike basalts their freshly broken thin edges are translucent when held in bright light. The phenocrysts in andesites commonly consist of striated feldspar and one or more dark minerals (hornblende, pyroxene, or biotite). Quartz phenocrysts are absent (distinction from rhyolite). Andesite was named for the Andes Mountains, where it occurs in great quantity and in many varieties.

Hornblende and Granite (Photomicrograph)

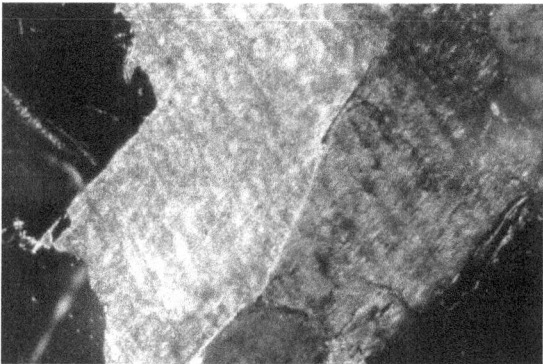

Biotite and Plagioclase (Photomicrograph)

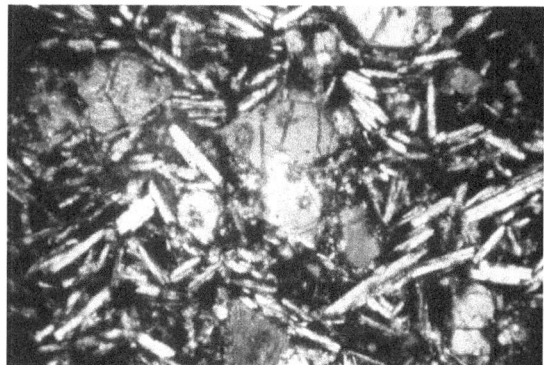

Hornfels and Olivine Basalt (Photomicrograph)

Felsite. The difficulty of discriminating between rhyolites and andesites that are devoid of phenocrysts makes it necessary to use an elastic noncommittal name. For the light-colored rocks of this kind, namely those which are white, light to medium gray, light pink to dark red, pale yellow, brown, purple, or light green, the term felsite is convenient. Some felsites that in hand specimens are as dark as basalts should be examined on their thin edges, where the rock will be seen to be almost white in transmitted light. Basalt specimens are dark even on thin edges.

Basalt. Lavas that are dark gray, dark green, brown, or black are termed basalt, the common extrusive equivalent of the basic magmas. Basalts are either compact or vesicular. If the vesicles have become Filled with some mineral, such as calcite, chlorite, or quartz, the fillings are called amygdales and the rock is an amygdaloidal basalt. Many basalts have no phenocrysts but others contain abundant conspicuous phenocrysts, consisting of feldspar, olivine, or pyroxene, or some combination of these. Therefore, in the Table of Igneous Rocks, basalt is shown to fall in both Classes III and IV. The phenocrysts are hard and have straight clean-cut boundaries, whereas amygdales are generally soft and have irregular, roundish or elliptical boundaries. Dolerite is the name given to those coarser-grained basalts in which the grains are large enough so that the constituent minerals are nearly or quite recognizable. There is no hard and fast line between basalt and dolerite on the one hand and dolerite and gabbro on the other.

Glassy Rocks

Volcanic glasses occur as thin crusts on the surfaces of lava flows, or more rarely as lava flows that have cooled rapidly. Most glasses are the products of the chilling of silicic magmas. Brilliantly lustrous volcanic glass is called obsidian and the duller and more pitchy variety is pitchstone Pumice is frothed glass. Obsidians are generally dark colored to black, yet many of them have the same chemical composition as rhyolite and granite. Hence obsidians seem to contradict the rule that nearly all silicic rocks are light colored. However, if the thin edge of a piece of black obsidian is held to the light, it will be seen that the obsidian transmits light and has lost much of its dark appearance. The deep coloring is the result of the uniform distribution throughout the glass of a relatively small amount of dark material. Basalt glass is of rare occurrence. To form it requires extremely rapid chilling of basaltic magma.

Summary of Igneous Rocks

There are three principal king of magma, termed, in the order of decreasing silica content: silicic, inter- mediate (mesosilicic), and subsilicic. Silicic magma forms, according to the environment in which it solidifies, granite, granite porphyry, and rhyolite; intermediate magma forms diorite, diorite porphyry, and andesite and subsilicic magma forms gabbro, dolerite, and basalt. These nine rock names constitute an irreducible minimum for the understanding of igneous geology, and can well serve as starting points for more detailed classification as the occasion may demand. The intrusive rocks are typically coarse grained: they are granite, diorite, and gabbro, comprising the so-called plutonic rocks. The extrusive equivalents of these plutonic rocks are rhyolite, andesite and basalt. Typically, the extrusive rocks are porphyritic, having conspicuous crystals that are set in a groundmass that is so fine grained as to be irremovable by the unaided eye. Basaltic magmas, however crystalline so readily, forming coarse-grained rocks, that the distinction between plutonic and extrusive equivalents is generally much less sharply marked than in the rocks formed from the more silicic magmas.

Sedimentary Rocks

Sedimentary rocks are formed principally in two ways. Some result from the accumulation of fragments derived from older rocks: detritus (pronounced detritus Latin deterere, detritus, to rub or wear away), consisting of particles of rocks and minerals is carried away from its source by water, wind or ice, and is eventually dropped by the carrier as sediment, and later this sediment hardens into rock. Detrital rocks thus formed are classified according to the size of their constituent grains of detritus. The second principal class of sedimentary rocks is made of material that was formerly dissolved in the sea, less commonly in is lakes from which it has separated either as the shells of organisms or as chemical precipitates; rocks made in this way are classified according to composition. By far the most abundant sedimentary rocks are shale sandstone, limestone, dolomite, and conglomerate.

Description

Conglomerate consists of gravel that has become firmly cemented The stones in it are more or less round (Fig. 17), having become water worn by abrasion during stream transport or by buffeting by waves in the shore zone. They may consist of rocks of any kind or of mixtures of many kinds, but durable material l, such as quarts and quartzite, is more common. In sine, the detritus ranges from coarsest sand grains up to pebbles, cobbles, and boulders many feet in diameter. The interspaces between the stones are filled with sand grains and a cement, which may be silica, clay, calcium carbonate, or iron oxide. Breccia is like conglomerate, except that most of the stones, instead of being round, are angular, with sharp edges and unworn corners. There are varieties intermediate between Breccias and conglomerate, in which the stones are poorly rounded or subangular, blurring the distinction between the well-defined end members. Breccias, in brief, are coarse angular detrital rocks. those that are firmly cemented alluvia l-fan material l are termed fanglomerates (p. 188), and those of glacial l origin till changed into rock are tillites (p. 195).

Sandstone. Sandstones are widely distributed detrital rocks of finer grain-size than conglomerates. They are of many colors; gray, yellow buff and tawny), red, and brown are most common, but green and other tints occur. Sandstones consist of sand grains held together by a cement. Coarse sandstones grade by increasing grain-size into conglomerates; on the other hand, fine sandstones by decreasing grain-size grade into siltstones, the lower limit for sandstone being that at which the individual grains cannot be distinguished by the unaided eye. Most sandstones are made up of grains of quartz or other materials not easily destroyed by weathering and transport. A sandstone is strong and durable if its cement is abundant and durable If the pore spaces remain more or less incompletely filled with cement the sandstone is proportionally porous. As is true of conglomerates, sandstones may have very different cements A siliceous cement produces the strongest and most durable sandstone, hence those most desirable as building stones. Sandstones when fractured break around the grains instead of through them, because the grains are stronger than the cement; consequently the broken surfaces have a gritty feel. If, however, the cement is so strong that fracturing takes place across the grains instead of around them, the rock is termed quartzite. Arkose is a sandstone containing abundant feldspar (more than 25 per cent). The feldspar is recognizable by its cleavage planes which reflect flashes of light as the specimen is turned from side to side while being examined. If an arkose contains much coarse feldspar, it closely resembles a granite. The arkose can be distinguished from granite, however, by the fact that its quartz is in angular or subangular particles, instead of being molded around the feldspar as it is in granite. Some arkoses contain leaf remains or other fossils, an impossibility in a granite.

Siltstone . Silts tone is a silt that has been converted into rock. It is intermediate in grain-size between sandstone and shale.

Shale. Gray in various shades is probably the prevalent color of shale. Red and pink in many shades, black, brown, buff, and green are also common. Shale is a consolidated clay and has a characteristic flat-conchoidal or "shelly " fracture, which is parallel to the bedding. Shale is so fine grained as to seem homogeneous to the unaided eye. It is soft enough to be scratched with a knife; harder varieties are termed argillite. Typically, shales have a smooth and almost greasy feel; but a little sand if present will make them feel somewhat gritty. Between clays and shales there are all gradations, but typical shale unlike clay, is not plastic when mixed with water. Some consolidated clays lack the characteristic shaly fracture, and massive rocks of this kind are sometimes designated mudstones or, more logically, claystones All these rocks are much too fine grained for the component particles to be determined with the unaided eye, or even with the microscope. Their composition is now being determined by means of X-rays, and it is already established that the composition varies widely.

Limestone. Gray is the most common color of limestones, but many others are seen. Limestones to which plant or animal remains have contributed carbonaceous matter are almost or quite black. Limestones range from extremely fine-grained varieties, irresolvable by the unaided eye, to fragmental and granular varieties. The ultra fine grained are made up of chemically precipitated calcium carbonate, or of the shells of microscopic organisms, or of a mixture of the two. In many coarse-grained limestones, whole or fragmentary shells of megascopic organisms can readily be seen. If the shells predominate and are loosely cemented, the rock is termed Coquina. The coarser- grained limestones show on their freshly broken surfaces distinct cleavage planes of calcite.

Since limestones are composed chiefly or wholly of calcite, they have a hardness of 3, and hence can be scratched easily. Limestones are "soft " rocks. They effervesce vigorously on being moistened with dilute acid, owing to the copious liberation of carbon dioxide gas. Many limestones are impure, containing admixed clay or fine sand. As the amounts of these admixtures increase, the resulting varieties of limestone grade on the one hand into shale or mudstone, and on the other into sandstone. Chalk is a loosely coherent variety of limestone loosely coherent because weakly cemented. It is extremely fine grained and is white or creamy white.

Dolomite. Dolomite resembles limestone closely, but is slightly harder and does not effervesce with cold acid, except on scratched or bruised surfaces, i.e., where the rock has been powdered. The grain- size ranges from aphanitic to megascopically crystalline. Some dolomites are coarsely porous. Dolomite, like limestone, is a carbonate rock. Whereas limestone is composed of calcite (calcium carbonate CaCO2 the rock dolomite is made up of grains of the mineral dolomite, which is calcium- magnesium carbonate CaMg (CO,).

Metamorphic Rocks

A metamorphic rock is a previously existing rock that has acquired a new mineral composition or new structures, or commonly both. The change from the older rock to the new rock was affected by a process acting within the Earth's crust, termed metamorphism. Such metamorphism takes place in response to changes in the geologic environment to which the pre-existing rock was subjected. The resulting metamorphic rock may retain vestiges of the original characters of the rock from which it was derived, but commonly the changes have been so thorough that the original characters were obliterated and the product is to all appearance a new rock. Most metamorphic rocks have a more or less parallel arrangement of their component minerals. If some of these minerals are of flaky habit, the parallel arrangement confers on the rock the capacity to split readily parallel with the direction in which the flakes are oriented. This tendency of a rock to split parallel to s plane is termed foliation (Latin folium, a leaf) because the rock breaks into leaves or thin slabs. All rocks having a foliation are grouped together as foliates The notable foliation-making minerals are the micas (muscovite and biotite), chlorite, and to a lesser extent amphibole which because of its needle-like habit makes a less well-defined foliation. Metamorphic rocks are abundant, are of extraordinary variety, and comprise some of the most remarkable rocks in the crust Only the simpler and more abundant kinds are described here.

Description

Gneiss. Gneiss (pronounced nice) has an imperfect foliation and is generally coarse grained. Many gneisses have a streaky, roughly layered aspect owing to the alternation of lenses or layers of unlike mineral composition, e.g., white lenses of quartz and feldspars may alternate with thin layers or streaks of black mica. Most kinds of gneiss contain mica, whose flakes are in parallel arrangement. The gneiss splits parallel to the direction marked by the mica flakes Probably the commonest kind is mica gneiss, containing abundant black mica. In some varieties both black and white mica occur together. Gneiss containing prisms of hornblende in more or less parallel alignment is called hornblende gneiss. Many gneisses are obviously granites that have had a foliation impressed on them; they are there- fore called granite gneisses.

Schist. Schist differs from gneiss in having closely spaced, better- developed foliation planes, and as a result it splits readily into thin flaky slabs or plates There is no demarcation between gneisses, which are imperfect foliates, and schists, which are well-developed foliates. In schists the minerals are large enough to be recognized by the unaided eye, a feature that distinguishes them from the finer-grained foliates termed phyllites Schists are generally named for the mineral whose parallel alignment produces the foliation. Thus mica schist contains much mica (biotite, Muscovite or both). Chlorite is the foliation-making mineral in chlorite schist and hornblende in hornblende schist. As schists split parallel to the plane in which the foliation-making minerals are oriented, these minerals seem to make up most of the rock; only by examining a schist on cross-fracture, i.e., at right angles to its foliation can it be seen that the schist contains other minerals, generally quartz. Many schists have scattered through them large, conspicuous, well-shaped crystals, simulating the phenocrysts of igneous rocks. A deep-red garnet commonly occurs in this fashion in mica schists, and such rocks are called garnet-mica schists. Cruciform crystals of staurolite occur in this way; the well-known lucky stones of Virginia are such staurolite crystals that have weathered out of the inclosing schists.

Phyllite. A phyllite (pronounced fillite) is intermediate in appearance between a schist and a slate. It is finer grained than a schist, so that its constituent minerals are not discriminable by the unaided eye. It differs from a slate in having a higher, glossy luster. Some phyllites, otherwise much like slates in appearance, contain sporadic large well- shaped crystals of garnet and other minerals. Phyllites by increase of grain-size grade into schists on the one hand, and by decrease of grain-size grade into slates on the other.

Slate. Slate is so fine grained that no mineral grains can be seen. Most slates are blue-black, a shade so typical that it is called slate colored, but many are red, green, gray, or black. Slate splits with a foliation so well defined that it yields slabs having plane surfaces al- most as smooth as the cleavage planes of minerals; hence foliation of this kind is called slaty cleavage. In roofing slates this cleavage attains its acme and allows them to be split into plane-parallel slabs of any desired thinness. Slaty cleavage, most remarkably, is not related to the bedding of the slate in which it occurs; in places it is parallel to the bedding but in others it intersects the bedding at angles ranging up to 90" (p.263). Slates grade on the one hand into phyllites, and on the other into shales. The distinguishing differences from shales are as follows: Surfaces of shales are generally of dull luster, whereas slate has a considerable luster. Slate is on the average slightly harder than shale. Most slates ring when struck a light blow, and sonorousness is a time honored test indicating high quality in a roofing slate. Slates when split yield nearly plane surfaces, whereas shales have a semi-conchoidal or "shelly " fracture.

Marble. Marbles are commonly gray or nearly white, but have many other colors. Many are streaked or splotched irregularly Marble is the metamorphic equivalent of both limestone and dolomite. The ordinary variety is composed of calcite and therefore effervesces readily when touched with cold dilute acid, whereas dolomite marble effervesces only if the acid is applied to a fresh scratch. All varieties of marble are soft enough to be scratched easily (hardness about 8). The term marble is reserved in geologic usage for those metamorphosed limestones and dolomites that are visibly crystalline to the unaided eye. In commercial practice, however, any limestone or .marble that will take a polish is called a marble, and is generally endowed with an alluring trade name. Although marbles are metamorphic rocks, few of them are foliated.

Quartzite. Quartzite consists chiefly of quarts, and therefore has a hardness of 7. The grains of quartz of which it is composed are so firmly cemented that when the rock is fractured, the fractures pass through the grains not around them, as happens in sandstone. Most quartzites have been formed by the metamorphism of sand- stones, but some are sandstones that have become so firmly cemented by a quartz cement that they too fracture across the grains and hence they also are called quartzites. Quartzites of metamorphic origin no longer show the sand grains of which the sandstone was composed, and tend to have a glassy appearance. Like marbles, most quartzites are massive and do not have the foliation so characteristic of most metamorphic rocks.

Contact-Metamorphic Rocks. A rich and varied assemblage of metamorphic rocks occurs around the borders of igneous masses. Limestones thus altered by heat are especially noteworthy for the handsome suites of well-crystallized minerals that occur in them garnet, vesuvianite, and many others and are favorite hunting grounds for those who enjoy collecting minerals.

The Principles of Petrology
by G.W.Tyrrell (1926)

The Science Of Petrology

Petrology is the science of rocks, that is, of the more or less definite units of which the Earth is built. In the nature of things the study is limited to the materials of the accessible crust, although we have in meteorites samples of rocks which must be identical with, or analogous to, those composing the interior of the Earth. The science deals with the modes of occurrence and origin of rocks, and their relations to geological processes and history. Petrology is thus a fundamental part of geological science, dealing, as it does, with the materials the history of which it is the task of geology to decipher. Rocks can be studied in two ways: as units of the Earth's crust, and therefore as documents of Earth history, or as specimens with intrinsic characters. The study of rocks as specimens may properly be designated petrography. Petrology is, however, a broader term which connotes the philosophical side of the study of rocks, and includes both petrography and petrogenesis, the study of origins. Petrography comprises the purely descriptive part of the science from the chemical mineralogical, and textual points of view. As we must gain an exact knowledge of the units we have to deal with before we can study their broader relations to geological processes and their origins, petrography is a necessary pre-requisite to petrology , and must be carried on as far as possible by quantitative methods, as in other physical and chemical sciences. The term lithology is nearly synonymous with petrology , but is now seldom used. Etymologically, it means the science of stones; and, accordingly, there is a tendency, which should be encouraged, to use the term to indicate the study of stones in engineering, architecture, and in other fields of applied geology. It would be equally appropriate and correct to speak of the lithology of conglomerates and breccias, when dealing with the stones contained in these rocks. Broadly speaking, petrology is the application of the principles of physical chemistry to the study of naturally- occurring Earth materials, and may therefore be regarded as the natural history branch of physical chemistry. Viewed thus, petrology is again seen as a subdivision of geology. The exclusive study of petrography sometimes tends to obscure this relationship, anti hides the fact that petrology is intimately connected with a host of fascinating geological problems.

The Earth Zones

Regarded as a whole, the Earth is a sphere of unknown material surrounded by a number of thin envelopes. The inaccessible heavy interior is known as the barysphere. This is followed outwardly by the lithosphere, the thin, rocky crust of the Earth; then by a more or less continuous skin of water, the hydrosphere; and finally by the outermost envelope of gas and vapor, the atmosphere. Other zones have been distinguished and named for special purposes. The zone of igneous activity and lava formation, situated between the lithosphere and the barysphere, is the pyrosphere; and water has distinguished the living envelope which permeates the outermost zones, as the biosphere. A zone towards the base of the lithosphere which can sustain little or no stress has been called the asthenosphere (sphere of weakness) by Barrell; and the zone in which crustal movements originate has been named the tectonosphere by certain Continental geologists. The Barosphere - While the interior of the Earth, of course, is inaccessible to direct observation, many facts have been indirectly ascertained about its structure and composition. It must, for example be hot. Observations in wells, mines, and borings, show that there is a downward increase of temperature, which is variable in different parts of the Earth, and averages 1' C for every 31.7 meters of depth, or roughly 50' C per mile in Europe. In North America the temperature gradient is much gentler, only 1 C to 41.8 meters of depth, or 38' C per mile. If these gradients continue indefinitely, enormous temperatures must prevail in the interior of the Earth, but it is probable that the rate of in- crease falls off with depth. Similarly, very great pressures must occur, even at moderate depths within the crust. The barysphere must also be composed of heavy materials. The density of the Earth as a whole is 5.6, but the average density of known rocks of the lithosphere is only 2.7. Hence the average density of the barysphere must be somewhat greater than 5.6. Several considerations serve to show that the barysphere must be rigid, with a rigidity greater than that of the finest tool-steel. In the early days of geology it was believed that the thin solid crust rested or floated on a molten interior; but when attention was given to geophysical matters it was soon shown that, under these circumstances, the thin crust would experience great distortion in response to the attraction of the moon, and, furthermore, owing to infernal friction, rotation would long be maintained. A strong confirmation of the rigidity of the barysphere is afforded by the study of earthquake vibrations. A heavy shock, say, in New Zealand, is recorded by seismometers in Britain about 21 minutes later, the vibrations having travelled by a more or less direct path through the barysphere. This speed of wave propagation is consistent only with high rigidity in most of the interior of the Earth.

Composition of the Earth Shells

The Earth has been called a projectile of nickel-steel covered with a slaggy crust. It is probable that all the planets and planetoids of the solar system have essentially the same composition; hence the wandering fragments of planetary matter known as meteorites or shooting stars, which the Earth sweeps up as it revolves in its orbit, are of particular interest in connection with the present topic. Meteorites fall upon the Earth's surface in masses which vary in size from the finest dust to huge blocks weighing many tons. Meteorites are divided into three main groups which pass gradually one into the other: 1. Siderites. The iron meteorites, consisting almost entirely of iron alloyed with nickel. 1. Siderolites. Mixtures of nickel-iron and heavy basic silicates, such as olivine and pyroxene. 3. Aerolites. The stony meteorites, consisting almost entirely of heavy basic silicates, olivine and pyroxenes, and resembling some of the rarer and most basic types of terrestrial igneous rocks. There are small amounts of sulphur phosphorus, carbon, and other elements in meteorites, which, however, may be disregarded in the present connection. Professor J. W. Gregory shows that if all known meteorites are considered, the iron group far outweighs the stony group. The stony meteorites fall in greater abundance, but the siderites fall in such large masses that they bulk much greater than the aggregates of small aerolites Hence the relative masses of the different types of meteorites support the above-cited view of the composition of the Earth. From geophysical data based on the distribution of density, earthquake vibrations, etc., Williamson and Adams have arrived at the conception of Earth composition illustrated by Fig. 1a. In this the Earth is shown to be built of four layers: (1) a thin surface crust of light silicates and silica; (2) a zone of heavy silicates (peridotite) which, of density 3.3 in its uppermost layers, is of density 4.35 in its lowest part at a depth of 1600 kms.; (3) increasing admixture of nickel-iron leads to a zone consisting of material similar to siderolites (pallasite) which, with a rapidly-mounting pro- portion of nickel-iron, passes into; (4) the purely metallic core. The actual composition of meteorites, as given above, supports this hypothesis. According to this view the Earth is conceived as the result of a gigantic metallurgical operation analogous to the smelting of iron, with the more or Jess complete separation of metal and silicate slag. Professor V. M. Goldschmidt has put forward a view of Earth composition which differs from the above chiefly in the intercalation of a zone of metallic sulphides and oxides between the nickel-iron core and the shell of heavy compressed silicates. In this conception the analogy of copper smelting is kept in mind, involving the separation of metal. sulphide-matte, and slag.

Chemical Composition of the Crust

The outer crust of the Earth down to a depth of 10 miles or so, consists of igneous rocks and metamorphic rocks, with a thin, interrupted mantle of sedimentary rocks resting on them. According to Clarke and Washington the lithosphere down to a depth of 10 miles is made up of igneous rocks, 95 per cent.; shale, 4 per cent.; sandstone, .75 per cent.; and limestone, 0.25 per cent. The metamorphic rocks, which are derived from igneous and sedimentary rocks by alteration under heat and pressure, are regarded as belonging to their initial types, and are neglected in the following calculation. The average chemical composition of each class of rock has been obtained by computation from a large number of existing analyses; and if these are combined in the proportions given above the following figures are obtained, representing the composition of the lithosphere down to the 10-mile limit of- depth. Column 2 shows the composition as recast into the form of oxides: On the assumption that the 10-mile crust entirely is composed entirely of igneous rocks, the results of the computations only differ on the whole, from those given above, in the second figure of decimals. It thus appears that fifteen elements between them make up 99.75 per cent of the Earth's crust, and that the majority of the elements which are important in human affairs are included in the remainder, being present in the crust in amounts of the order of 1/100th of 1 per cent. It has been asserted that the Clarke-Washington method seriously over-estimates the acidity {amount of silica) of the crust, since no allowance is made for the relative amounts of the different kinds of igneous rock composing the average. Rocks of acid composition are largely restricted to the continents, and are probably underlain at shallow depths by basic rock (basalt, etc.). The floors of the oceans also probably consist of basalt. Hence, perhaps, the average crustal rock should be regarded as considerably more basic than is allowed by the Clarke method. Rocks have already been defined as the more or less definite units constituting the Earth's crust, but in popular usage the term rock denotes any hard, solid material derived from the Earth. In geology, however, the term is often used without reference to the hardness, or state of cohesion, of the material; and sand, clay, or peat, are thus just as much rocks in the scientific sense as granite and limestone. When solid rocks are examined closely they are found to consist largely of fragments of simpler chemical composition, which are called minerals or strictly, mineral species. Hence we arrive at another definition of rocks as aggregates of mineral particles. It is important to distinguish the two uses of the term mineral It is used in a perfectly Legitimate sense by the ordinary man, the miner, prospector, quarry- master, lawyer, landowner, and the scientific mineralogist, to indicate materials, such as coal slate, clay, etc., which are won from the Earth's crust, but which, in the strict petrological sense, are rocks. We speak also of "mineral waters " and " mineral oil " in the same way. Natural glasses and other amorphous materials which occur as rocks and in rocks may also be regarded as minerals in this sense. The materials of which rocks are largely composed are, however, mineral species i.e. natural inorganic substances with a definite chemical composition, or a definite range of chemical composition, and a regular internal molecular structure, which manifests itself under favorable circumstances by the assumption of regular crystalline form, and the possession of definite optical and other measurable properties.

Conglomerate and Arenite Sandstone (Sedimentary Rocks)

Gneiss and Chlorite Schist (Metamorphic Rocks)

Gabbro and Basalt (Igneous Rocks)

Rocks and Their Composition

The Rock Minerals

The principal chemical elements present in the Earth's crust combine with oxygen, the most abundant element, to form oxides; and this, by the way is the most convenient and usual method of presenting the chemical analyses of rocks. These elements however fall into various combinations from which the rock-forming minerals arise, in which oxides, as such, are only of secondary importance. Silicates are the most abundant compounds constituting the rock-forming mineral oxides come next; then carbonates, phosphates, sulphates, etc., in greatly diminished importance. Among the elements, only iron {in the basalt of Disko, Greenland), carbon {as diamond and graphite), and sulphur {volcanic action; decomposition of sulphates and sulphides, occur native. It is probable that 99.9 per cent of the Earth's crust is composed of only about twenty minerals out of the thousand or so which have been described. These are the rock-forming minerals par excellence. Referring first to the silicates, the feldspars are by far the most abundant and important group, not only of the silicate but of the rock-forming minerals in general. The chief members are orthoclase and microclines both silicates of potassium and aluminum and the various plagioclase which are mixtures in all proportions of the two end-members, albite, silicate of sodium and aluminum, and anorthite silicate of calcium and aluminum. Allied to the feldspars, but containing less silica in proportion to the bases present, are the felspathoid minerals, of which the most important are nepheline silicate of sodium and aluminum (corresponding to albite among the feldspars), and leucite silicate of potassium and aluminum (corresponding to orthoclase among the feldspars). The mineral analcite, a silicate of sodium and aluminum with combined water, properly belonging to the group of zeolites, may nevertheless take its place as a rock-forming mineral with the feldspathoids. The mica group forms a link between the alkali-alumina silicates above mentioned and the heavier and darker Ferro-magnesia silicates to be described later. Of micas proper, the two chief members are the white mica, muscovite, silicate of potassium and aluminum with some hydroxyl, and the dark mica, biotite silicate of potassium, aluminum, magnesium, and iron, with hydroxyl. Chlorite is the green hydrated silicate of magnesium and iron, of micaceous affinities, which is the most familiar alteration product of biotite and other ferro-magnesian minerals. Among the ferro-magnesian minerals proper there are the three main groups of the pyroxenes, amphiboles, and olivine's. The pyroxenes are metasilicates of calcium, magnesium, and iron, of which the two chief members are the orthopyroxenes (enstatite and hypersthene), simple metasilicates of magnesium and iron; and augite, the monoclinic pyroxene, a complex metasilicate of calcium, magnesium, iron, and aluminum. The amphiboles form a parallel group to the pyroxenes, but with different crystal habit. The chief member is hornblende, a mineral of composition similar to that of augite, but usually richer in calcium. The olivine's are simple orthosilicates of magnesium and iron, and stand in the same relation to pyroxenes and amphiboles as the feldspathoids do to the feldspars. Serpentine is the hydrated alteration product of olivine and other ferro-magnesian minerals. Numerous other silicates occur as rock-forming minerals, but it is only necessary here to mention the garnets, a varied isomorphous series, chiefly silicates of iron, magnesium, calcium, and aluminum; epidote, silicate of calcium, iron, and aluminum, an abundant alteration product of lime-rich silicate minerals; andalusite, kyanite, and silliminite, all simple silicates of aluminum; cordierite, silicate of magnesium, iron, and aluminum; and staurolite, silicate of iron and aluminum. The five last-named minerals are characteristic of the metamorphic group of rocks. Among the oxide minerals only four need be mentioned as prominent rock-formers. Quarts, the dioxide of silicon, is, perhaps, the most abundant mineral next to the feldspars. Impure colloidal silica, especially the dark variety known as chert, also forms large rock masses. The oxides of iron come next: magnetite (Fe_3O_4) is very widely distributed in rocks in small quantities; hematite (Fe_2O_3) and limonite (Fe_2O_3), nH_2O) form the universal red, brown, and yellow coloring matters of rocks. Ilmenite, the oxide of iron and titanium $(FeTi)_2O_3$, is perhaps even more widely distributed in rocks than magnetite. Of carbonates, the minerals calcite, the carbonate of calcium, and dolomite, the carbonate of calcium and magnesium, are by far the most abundant, and are the chief minerals of the important group of the limestone rocks. One phosphate, the mineral apatite, phosphate of calcium with some combined fluorine or chlorine is universally distributed in small amounts; and one sulphide, the ubiquitous barites, disulphide of iron, is a common rock-forming mineral. Two sulphates, gypsum sulphate of calcium with combined water, and barites, sulphate of barium, occasionally form rock masses; as also one chloride common salt or halite the chloride of sodium. For information regarding the crystallographic optical and other properties of the rock-forming minerals the reader is referred to one of the numerous standard textbooks on that subject.

The Classification of Rocks

Whatever theory of Earth origin be held it is at least certain that all parts of the original surface of the Earth passed through a molten stage, and that the first solid material which existed was derived from a melt or magma. This original crust is nowhere exposed on the present surface, but all subsequently-formed rocks, in the first instance, have been produced either from this, or from later irruptions of molten matter. Rocks formed by the consolidating of molten magma are said to be Primary or Igneous. After the solidification of the original crust, and the formation of the hydrosphere and atmosphere, the waters and the air, both probably of much greater chemical potency than now, began, to attack the primary rocks. Disintegrative action produced loose debris, and chemical action produced both debris and material in solution. The loose fragments would be swept away by water and wind and would ultimately collect in the hollows of the crust, where also the waters and soluble matters would be found. The collected debris deposited from suspension in water or air would finally be cemented into hard rock, and would be thus added to the solid crust. Under suitable circumstances the soluble matter likewise would be precipitated, either directly, or indirectly through the agency of organisms, the latter, of course, in somewhat later geological times. The rocks thus produced would eventually become solid and help to build up the crust. These processes have gone on through- out geological time, the newer increments of the crust under- going attack as well as the older parts. Hence it may be that some of the material has gone through many successive cycles of change. The rocks formed in these ways are called Secondary, because they are composed of second-hand or derived materials. They may be divided into Sedimentary, Chemical, or Organic according to the process by which they received their most distinctive characters. Finally, both Primary and Secondary rocks may be subjected to Earth movements which carry them down to dept is in the crust where they are acted upon by great heat and pressure. By these agencies the rocks are partly or wholly reconstituted; their original characters are partly or wholly obliterated, and new ones impressed upon them. Rocks thus more or less completely changed from their original condition are known as the Metamorphic rocks. We thus arrive at the time honored three-fold classification of rocks according to their modes of origin into Igneous, Secondary (Sedimentary), and Metamorphic. The Primary rocks are distinguished by the presence of crystalline minerals which interlock one with the other, or are set in a minutely- crystalline paste, or in a glass. They show signs, as do present-day lavas of having cooled from a high temperature. They are usually massive, unstratified unfossiliferous and often occupy veins and fissures breaking across other rocks, which they have obviously heated, baked, and altered. The Secondary rocks are composed of clastic and precipitated materials, or of substances of organic character and origin. The materials are often loose and unconsolidated, or are welded together by pressure or by a cementing sub- stance into a solid rock. They are further distinguished by the frequent presence of bedding or stratification organic remains (fossils), and other marks indicative of deposition from water or air in the sea or on land. The metamorphic rocks present characters which, in some respects, are intermediate between those of the Primary and Secondary rocks. Great heat and pressure cause recrystallization; hence, like the Primary rocks, they often consist of interlocking crystals Furthermore, pressure causes the development of more or less regular layers, foliation or banding, in which the Metamorphic rocks resemble those of Secondary origin. Since the Metamorphic rocks are formed from pre- existing igneous or sedimentary rocks they often retain traces of their original structures. Mr. T. Crook has recently formulated a genetic classification, in which the rocks arranged according to a geological grouping of processes. He divides rocks into two great classes: 1. Endogenetic formed by processes of internal origin which operate deep seatedly or from within outwards (with respect to the crust). High temperature effects are predominant, and the water associated with the processes is partly of magmatic origin. 2. Exogenetic formed by processes of external origin, operating superficially or from without inwards. These rocks are formed at ordinary temperatures, and the associated water is of atmospheric origin. In the Endogenetic class are included the igneous rocks (along with certain pneumatolytic and hydrothermal types), and metamorphic rocks; and in the Exogenetic class come the rocks usually classed as sedimentary. The following gives a simplified statement of the classification l. Endogenetic Rocks. (1) Igneous Rocks. (2) Igneous Exudation Products (due to pneumatolysis, metasomatism, etc.). (3) Thermodynamically-altered Rocks {Metamorphic Rocks). 2. Exogenetic Rocks. (1)

Weathering Residues. (2) Detrital Sediments. (3) Solution Deposits. (4) Organic Accumulations

Granite

By Rob Kanen

Micrograph and Specimen of Granite

Origin of Granite

Granitic magma is a general term used to describe magma that is similar in composition to granite; that is, containing greater than 10% of quartz. Outcrop of plutonic granite on the Earth's surface requires some kind of erosion to expose the buried granite. Granites may take the form of batholiths; sills and sheets; swarms of plutonic intrusions or migmatite complexes. They form the major part of surface exposure of continental crust. Plutonic rocks also exist in the oceanic crust; however, by geophysical methods and drilling, these have been determined to predominantly be of basic or ultrabasic composition. Granites are associated with volcanic areas, continental shields and orogenic belts. To explain their emplacement, it is first necessary to obtain an understanding of their origin. The study of outcrops, geophysical surveys and of its extrusive equivalent, rhyolite, are some of the original methods used. In general, two feasible theories resulted. One, known as the magmatic theory, states that granite is derived by the crystal fractionation of magma. The second, known as the granitization theory states that granite is formed "in situ" by ultrametamorphism. There is evidence to support both theories and current thinking is that magma forms from both processes; in many instances, from a combination of the two.

Introduction to Geology

The magmatic theory involves the use of the Bowen Reaction Series. Thus, if crystal fractionation of a magma of tholeiitic basalt composition were to occur, one of its end products would be granite. In many places, emplacement of granite plutons is synchronous to volcanic eruptions. They commonly form ring complexes around 10 km in diameter with volcanic remnants that have subsided into the couldron as central blocks. This has occurred in the Permian Oslo Graben Province (Carmichael et al., 1974). They also occur as stocks and plutonic intrusions near volcanic centers which consist of granodiorite in an andesitic volcanic province. The former are generally alkali granites in a rhyolitic province. Such plutons typically show sharp contacts; a lack of deformation in the country rock, chilled margins and contact aureoles. All of these phenomena suggest that granite was emplaced as a liquidus magma. Chemically, there is similarity in the composition of many granite plutons to their extrusive associates, the andesite-dacite-rhyolite series of rocks. This suggests there is some kind of relationship between the emplacement of granite plutons and volcanism. Some of the theories that have been postulated include that granite is a surface expression of magma derived from a deep seated batholiths and that plutons are derived from the same source as the extrusive but along separate supply routes. Gilluly (1963) has found that volcanism in the Western United States has been more prolific, continuous and diversified than granite plutonism. He found that at some areas and at certain periods, volcanic products were mainly andesitic, at other times and another place, they were mainly basaltic and so on. All types were very prolific. In terms of plutonic rocks, quartz diorites, granodiorites and quartz monzonites were dominant. He concluded plutonism must depend on processes whose time and scale is of essentially a different order of magnitude from those of volcanism and tectonism. He also suggests that the volcanic magmas have been derived from the oceanic crust and plutonic magmas from more siliceous materials.

An interesting question in magmatic evolution of granite batholiths is how is the immense amount of country rock removed to make room for the batholith? Various mechanisms have been proposed: massive explosions removing the country rock; formation of ring dykes and couldron subsidence; lifting of the country rock and subsequent erosion; and digesting of the country rock as the magma rises. Not one of these would adequately explain its removal on its own but rather it is more likely that all of these occur at some time or other.

Introduction to Geology

The granitization theory explains the origin of granite by the process of ultrametamorphism or anatexis. Anatexis is defined as the melting of pre-existing rock to give granite. The "crux" of the granitization theory is migmatite. Migmatite consist of two components: one light colored granitic component, called neosome, and a dark metamorphic component called paleosome. Both components have been ultra-mixed. In the granitization theory it was thought that migmatites were rocks in the process of becoming granite. Thus, the neosome, granitic component, was anatexic component; in which case, it may have formed by partial melting of the rock in place and segregation of the melt from the solid, or migration of the melt from its source of origin and intrusion into the host rock. The resultant migmatite which formed by the first process is called a venite, the migmatite formed by the second process is an orterite. Mehnert (1963) reinforces this idea. He has mapped granodioritic masses up to 10km in diameter with a homogenous center and increasingly heterogeneous material towards the outer border. There is no distinct contact of the granodiorite with the country rock, but rather grading into a migmatite zone and finally a gneiss containing eyes of oligoclase and potash feldspar.. He interprets this as meta-greywacke which has been completely fused in the center, granodiorite zone, and partly fused in the migmatite zone. He proposes the outer, gneissic zone, is metamorphic in origin, but due to metasomatism rather than fusion of a melt. In more recent years, it has been recognized that migmatite can form in various other ways. These are granitization, by ion exchange and diffusion, especially with K+ and Na+; mobilization and injection of granitic material from depth, an; and alkali metasomatism, using aqueous pore solutions as a medium. Thus, granites can form both by magmatism and granitization or a combination of the two.

In recent times, it has been shown that the relationship between regional metamorphism and the formation of granite (ultrametamorphism) is more complex than originally thought. Autron et. al.(1970) shows that the formation of large volumes of granite have formed during periods where there has been no metamorphism; for example, the Lower Silurian Caledonian Granites. However, crustal thickening was occurring and it is thought that this takes place as a consequence of crustal shortening by collision and under plating by subduction. The local thickening of the crust provides a sufficient rise in temperature at depth for crustal melting to occur. This may rise in several different ways to produce granite. Experimentally, it has been proven that granite can form by remelting of the crust; however, it has also been shown that the temperatures and water concentrations at depth might not be high enough for remelting to occur (Brown and Hennessy, 1978) unless magma from the mantle influenced in some way.

When this theory is applied to subduction zone areas, remelting of the underplated crust will not occur because of a lack of heat and volatiles. In terms of a collision type situation between two plates, it is thought sufficient water is available from the dehydration of micas during metamorphism of the metasedimentary crustal wedges for remelting to occur. Lochenbruch(1968) suggests that although this may occur, it is impossible for granitic melts of such origin to be homogenous. However, other authors have argued that homogenization can occur fairly easily (e.g. Talbot, 1971).

An interesting problem in an anatexic origin is explaining how the granite melt separates from its associated solid material, especially in pseudo-plastic, or Bingham bodies. Weertman (1971) proposed shear melting as an important part of the mechanism. Various mechanisms to explain how magma rises through the crust have been discussed and proven acceptable. These include gas fluxion, tectonic squeezing, expansion on melting, seismic pumping and diapirism. All of them are related to the moving of magma down a pressure gradient. Opposing the upward driving force of these mechanisms is drag. Drag increases rapidly as the magma cools, especially when it collects large amounts of xenoliths. Thus, different parts of the magma body ma move upwards at different rates. This provides another explanation for the lack of compositional homogeneity and difference in internal contacts.

Magmas make their way up through the crust via major lineaments, such as fault zones. This is very evident in the Andes and the Coastal Batholith of Peru (Pitcher and Bussel, 1977). Leake (1978) suggested that not only do lineaments provide a route for magmas but also generate them where deep faults cause large pressure reductions and shear zones, where segregation of crystal mushes and Bingham Bodies could occur. It is thought that formation of massive granite bodies via lineaments is restricted to fractured continental plate edges, where massive lineaments are present. In areas which have not undergone large scale faulting, intrusions are more dispersed.

Estimating the depth at which a magma is emplaced is very difficult. However, a very rough approximation could probably be obtained by the fluid inclusions in minerals; the composition of the minerals; for example the aluminum content of hornblende and the ordering of alkali feldspars. However, this is restricted, as feldspars may have reacted to weathering and metamorphism. Furthermore, in the formation of depth models, adjacent intrusions are inferred to be syndepositional in origin and to have a vertical continuity, for example, Buddingtons epi-, meso- and kato-zones. Another depth model recognized by Eskola(1938) and Read (1950) proposes a granite series showing different features at different depths. The first zone, zone of differential anatexis, is the lowest level in the crust. Some granitic magma is formed in situ and mobilization begins, venites are also formed. The zone of injection, or potash metasomatism, is the zone in which the crystal mush and anterites form. Portions of the magma become more liquid and rise to its upper surface. Metasomatism also occurs, indicated by large k-feldspar crystals. In the upper zone, regional metamorphism precedes or accompanies emplacement of the magma. Many veins and sharp contacts indicate the fluid nature of the granite. Fyfe(1970) envisages a tear dropped shaped magma melt with tail inverted rising upwards by diapirism through a cooler, more solid country rock which may have left on it an imprint, in the form of migmatite. "..some migmatite zones might represent a region through which a drop [evolving pluton] passed rather than a region where melting started"

According to Pitcher (1979) granite magma may be emplaced forcefully or permissively into the upper crust. During forceful intrusions, the pushing aside and updoming of country rock occurs. In some cases, they have been found to their way through the crust, especially where there is little overburden. Here, there is little or no evidence of strain in the immediate country rock. In deeper environments, radial distension occurs. There is often little evidence of drag occurring but radial shortening as indicated by strain markers, clasts in the country rock and xenoliths in the intrusive. Pitcher (1979) says, "that in examples of expanding diapirs in Donegal the superimposed deformation aureoles extending almost exactly as far as the thermal recrystallization aureoles, indicating that plastic deformation depended upon the pre-heating of the envelope." He goes on to say, " the succession of metasediments entering the aureole of the Ardara multiple diapir [at Donegal] is thinned to at least 10% of the original thickness, and although the exact deformational path which leads to this result cannot be deduced from the presently observed fabric (Dixon, 1975), Holder (1978) showed that these are incremental increases in strain in the aureole rocks of the Ardara Pluton which correlate both with the several phases of metamorphism and the intrusion of magmatic pulses." In other words, evidence for the occurrence of radial distension is very strong, forceful intrusion (forceful punching) is on a much weaker footing. It may also explain why there is no evidence of the envelope surrounding the intrusion, flowing around and behind it. Furthermore, new pulses of magma into the pluton, expanding and contracting it, suggests plutons are continuous tubes connected to the source, not individual drops (Pitcher, 1979).

Mechanisms for the permissive intrusion of granitic magmas include subsidence, couldron upheaval and stoping. These are characteristic of shallow seated, sub volcanic or aureole type granite plutons. The upward thrust combined with thermal expansion upon cooling can still result in updoming and fracturing of the overlying material. Fracturing of the country rock to allow intrusion of magma is thought to be mainly due to thermal expansion. However, new fractures may also form from the seismic shocks which accompany the emplacement of magma. Hydraulic fracturing could extend the initial fractures. Myers (1975) contends "…that gas penetration along micro-breccia zones entrains the fragments, forming a proto-tuffite, so opening a pathway for the advancing gas charged magmas". This allows easy penetration of the country rock and the prizing (stoping) off of xenoliths. A lot of the supportive evidence for this process is in the form of ring dykes containing large amounts of xenoliths. Widening of the channels by gas clearing and gas entrainment may also occur in a vertical sense, thereby removing country rock up or down. Such a mechanism is necessary to explain the removal of the country rock, especially since ring dykes are vertical or dipping inward.

Removal of country rock by subsidence is a proven fact. In some places the degree of subsidence can be measured. Nested or stacked intrusions have been explained by the subsidence of a central block in a ring dyke and infilling of the space left above it by pulses of magma. A widely accepted theory is that they form by magma being pumped into a pluton which has a solid outer surface, but still crystallizing inner core.

Whether the couldron or diapir will form depends upon the contrast in ductility between the country rock and the magma. Pitcher (1979) says it is wrong to assume that the ductility contrast decreases with depth. It depends on other factors also, such as, deformation and metamorphism of the crust. As deformation increases, ductility decreases, therefore, the early intrusions are more likely to be diapiric and the older intrusions in the form of a couldron. However, diapirs have also formed at the end of a sequence, therefore, low viscosity, liquid magmas may be injected to form couldrons and higher viscosity, crystal mushes may form diapirs.

In terms of classifying granitic magmas according to their origin, Pitchers (1979) classification scheme is probably best. In terms of orogenic environments, he recognized a crustal S-type produced by anatexis and which is compositionally restricted; and a mantle I-type derived from a basic igneous source and which is compositionally expanded. To differentiate between the two, measurement of the Sr87/Sr86 ratio is done. Those that are higher than 0.7060 tend to be S-types, those lower, I-types.

The origin of emplaced granitic magma is diverse and mechanisms used to explain its emplacement varied. As techniques for studying granite become more refined, a greater understanding of them will result.

Sedimentary (S-) and Igneous (I-) Type Granites

Classification of granites according their magmatic origin results in the formation of two contrasting groups, S-types and I-types. S-types result from the partial melting of metasedimentary source rocks, a process called anatexis or ultrametamorphism. I-types are derived from source rocks of igneous composition that have not gone through the surface weathering process, or from crystal fractionation of magmas. It is more often than not very difficult to determine whether a granite has a particular origin. As a result, many techniques have been developed to determine whether a granite is an S-type or an I-type. Some of these techniques are very simple, only requiring a study of mineral assemblages and inclusions; however, to determine beyond doubt whether a granite is of a particular type a combination of different analysis must be used.

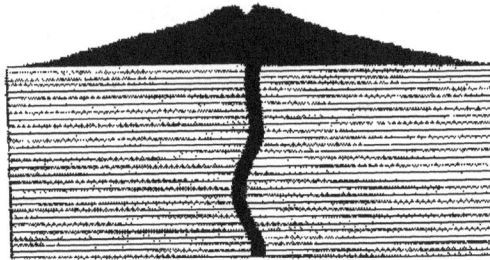

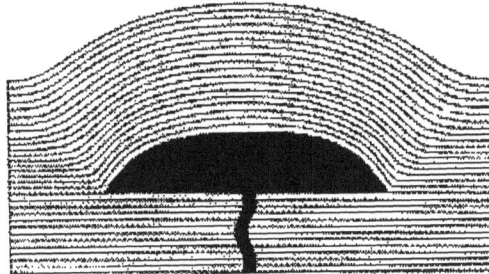

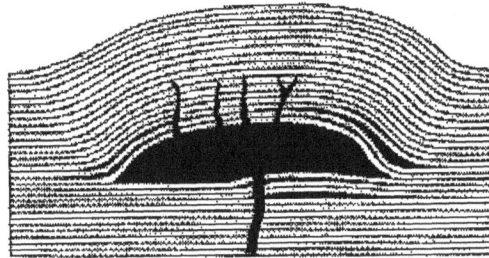

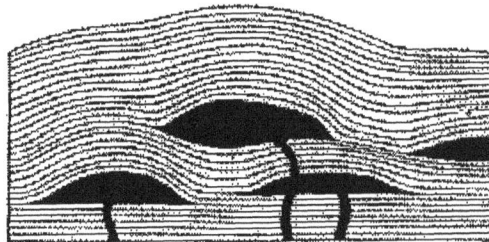

Emplacement of Magma

S- and I-type granites from the Tasman Orogenic Zone of Eastern Australia were studied in detail by Chappell and White (1974) and others. The granites, which are of widespread occurrence, may be distinguished by chemical, mineralogical, field and other criteria. White, Chappell and their coworkers (1974, 1978, Hine et al, 1978) have carried out a complete study of magma provenance in this area, particularly in the Lachlan Fold Belt. They recognized a group of early, metamorphically harmonious plutons largely composed of S-type granites, which probably originated from the remelting of metasediments, and a younger group of mostly I-type granites with aureoles derived from remelting of deep seated igneous material.

Geochemical Methods

It has been shown that the crustal S-type granites are compositionally restricted and the mantle I-type granites are compositionally expanded (Chappell and White, 1974). This is reflected by marked differences in geochemical parameters. Chappell and White (1974) used a number of chemical properties to distinguish between S- and I-type granites. Hine et al (1978) uses a more detailed chemical analysis to confirm and expand upon Chappell and Whites findings.

Chappell and White distinguish S-type and I-type granites using numerous chemical parameters. I-types have relatively high sodium, Na2O greater than 3.2% , in felsic varieties, decreasing to more than 2.2% in mafic types. S-types have relatively low sodium, Na2O normally less than 3.2% in rocks with approximately 5%K2O, decreasing to less than 2.2% in rocks with approximately 2% K2O. Furthermore, S-types have been determined to have a Mol Al2O3/(Na2O+K2O+CaO) ratio of greater than 1.1 and I-types less than 1.1. Another distinctive chemical property determined by Chappell and White is the normative corundum in S-type granites, being greater than 1% CIPW corundum. In I-type granites, less than 1% CIPW diopside is present. I-type granites have regular inter-element variations within plutons with linear or near linear variation diagrams. The variation diagrams of S-type granites is more irregular.

These chemical properties result from the removal of sodium into sea water, or evaporites and calcium into carbonates, during sedimentary fractionation (weathering). Subsequent, relative enrichment of the sediment in aluminum must have occurred.. S-type granites com from a source which has been subjected to this sedimentary fractionation.

Hine et al (1978) used the Kosciusko Batholith to illustrate chemical differences between S-type and I-type granites. A Na2O/K2O plot of the Kosciusko batholith illustrates the fundamental chemical difference. The more potassium rich S-types are lower in sodium, a very distinctive phenomenon. Differences in this type are useful in recognizing S- and I-type granitoids and were an important factor in deciding what the overall S- and I-type characteristics were inherited from different sources (Chappell and White, 1974). The formation of shales by chemical weathering processes enriches Al relative to Na and Ca, since Na is removed into sea water and Ca is concentrated in carbonates. Pelitic rocks have high K/(Na+Ca) ratios and this is reflected in the high K/Na ratios of S-type granites and their relatively low Ca contents.. They directly result in the higher Al/(Na+K+Ca/2) ratio of S-type granites.

Hine also used oxidation states of iron in his attempt to classify the Kosciusko Batholith granitoids. Flood and Shaw (1975) suggest the presence of carbon or sulphur in the sedimentary source rocks results in S-types being much more reduced than I-types. This is consistent with the Kosciusko Batholith rocks studied by Hine et al.

Chappell and White suggest S-type granites are restricted in composition to high SiO2 types, while I-type granites have a wide composition from felsic to mafic. These characteristics are a consequence of S-type granitoids having been derived from a more SiO2 rich source. Therefore, granitoids containing less than 655% SiO2 can generally be assumed to be I-type. This is consistent with the work done by Hine et al.

Another important geochemical feature of S- and I-type granites is the various isotope compositions. The initial Sr87/Sr86 ratios being higher in S-types because have been through and earlier sedimentary cycle.. A critical boundary of 0.7060 is suggested by Kistler(1974) and Armstrong et al (1977). Chappell et al (1974) suggest a boundary of 0.708. I-type granites have initial Sr87/Sr86 ratios between 0.704 and 0.706 (Chappell and White, 1974). Isochrons of I-types give a regular linear set of points whereas those of S-types show a scatter of points within a broad envelope, reflecting variations in initial Sr87/Sr86 within a single pluton as a result of a heterogeneous source material. Classification based Sr87/Sr86 ratios are valid.

O'Neil et al (1977) and Taylor (1977) have illustrated the importance of oxygen isotope ratios, and to a lesser extent, hydrogen isotope ratios in distinguishing between S- and I-type granites. In general, sedimentary rocks are much richer in O18 than primitive igneous rocks, therefore, if the granitoids have retained the characteristics of their presumed source material, O18 composition should be a useful tool for differentiating between S- and I-type granitoids.

O'Neill et al, studying the Berridale Batholith, determined that oxygen isotope compositions of whole rock samples are an excellent discriminant between the two types. O'Neill determined the average composition of O18 for S- and I-type granites of the Berridale Batholith to 9.9 to 10.5 and 7.9 to 9.4 respectively. A drawback in his argument is that he assumes the oxygen isotope compositions to be a remnant of the original material from which was partially melted to form the granitoid. Earlier, he says, "… the effect of alteration is to lower the O18 content of the rock. If this alteration took place in the presence of aqueous fluid that was in near oxygen isotope equilibrium with these cooling plutons (i.e., a magma derived deuteric fluid), the lower temperature chloritization process should have increased the O18 content of the rock". Clearly, the use of O18 compositions to classify granitoids is restricted to unaltered terrain's. In cases where whole rock analysis does not clarify the problem, O'Neil says the analysis of mineral separates can often provide answers.

Discrimination based on hydrogen isotope composition is less precise than using oxygen isotopes, but average hydrogen isotope compositions lower than -80 almost certainly indicate I-types. S-types have average values clustering around -62.

Petrographic Methods

Some of the geochemical aspects of S- and I-type granitoids are reflected in the mineralogy. Hornblende is common in the more mafic I-types and is generally present in the felsic types also. In the felsic S-types, hornblende is absent, but muscovite is common, while in the mafic

S-types, biotite is often very abundant. Monazite is the usual accessory in the S-types whereas sphene is common in the I-types. Garnet and cordierite may occur in S-type xenoliths as well as in the granites themselves. Apatite inclusions are common in biotite and hornblende of I-types, but occurs in larger individual crystal in S-types. Thus, I-types characteristically contain biotite+hornblende plus/minus sphene plus/minus monazite. S-types contain biotite plus/minus muscovite plus/minus cordierite plus/minus garnet plus/minus ilmenite plus/minus monazite.

A fairly detailed petrological analysis of S- and I-type granites was carried out by Hine et al (1978) of the Kosciusko Batholith. Hine and his coworkers determined the S-type granitoids to be dominantly quartz rich adamellites and granodiorites with a few felsic granites and mafic tonalites. A modal distribution shows the composition in S-types to be dissimilar to I-types, with S-types falling in the granite/adamellite range and I-types in the granodiorite/tonalite/monzodiorite range.

Field Methods

Observable features in the field may at times be distinctive. Chappell and White (1974) determined the more mafic I-types to contain mafic hornblende bearing xenoliths of igneous appearance whereas hornblende bearing xenoliths are rare in S-types. Metasedimentary xenoliths are common in S-types. Chappell and White observed that S-types are usually early in the intrusive sequence and often have a strong secondary foliation. They apply the term "restite" to relict, or residual, source material. They go on to say large, restite milky quartz inclusions are common in S-type granites. Restite occurs as xenoliths, clots or xenocrysts and may be used to distinguish between S- and I-type granites.

Conclusion

The methods that may be used to distinguish between S- and I-type granites include geochemical, petrographic and field methods. Not one single method may be used universally to make a clear distinction since outside influences such as post orogenic alteration and reaction with wall rock material may drastically change some of these criteria. In general, geochemical methods seem to be the most successful contrasting the differences between S- and I-type granites. Petrographic and field observations are extremely useful to supplement geochemical data

Volcanoes

By Rob Kanen

General

Volcanism is a term used to describe the process by which extrusive igneous rocks form. These rocks are named lava's. Lava's originate as a silica rich magma in the mantle of the Earth. Do to very high temperatures the magma is very fluid and since it has a low density it rises to the Earth's surface. Once the magma reaches the crust it may rise through cracks in the country rocks called fissures. As the magma rises up the fissures it leaves spaces behind which must be filled. The country rocks fills some of these spaces. As a result the country rocks fractures supplying additional areas of weakness where the magma may rise. The ejected magma solidifies and accumulates around the place it is ejected to form a dome like structure. The place of emergence of the magma is called the vent. Different types of magma may be ejected from the vent.

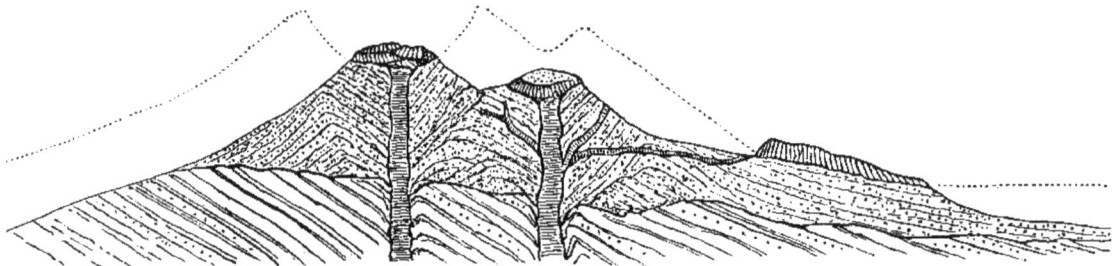

Structure of a Volcano

If the magma is of felsic composition it is called rhyolite, if it is of mafic composition it is called basalt and if it is of intermediate composition it is named andesite. Other substances may also be ejected from the event. Scoria is a solidified version of spatter, which is liquid drops of magma. If some of the solidified material surrounding the vent falls back into the vent it will be emitted as ash. Consolidated ash is called tuff. Tuff which has formed from falling ash is ash fall tuff and it has good size sorting. Tuff which has formed from ash flows or nue'e ardente explosions is called sorted. All material blown] out of the vent is called tephra. Tephra may be ash, fragments of rocks or volcanic bombs. Gasses may also be emitted. These include water vapor, CO_2, HCL and H_2S. The oxidation of H_2S produces sulfataras or sulfur deposits. Condensation of water vapor may cause ground water to heat up and form springs and geysers. In tropical climates where there is lots of moisture lahars or mudflows may occur.

Example of a Shield Volcano

The shape of the cone which forms around the vent of the volcano depends upon the viscosity of the magma and the topography of the land. The viscosity of the magma depends on the composition of the magma. Magma of felsic and intermediate composition is more viscous than magma of basic composition. Viscous magma which is of felsic composition does not flow very far before it solidifies. As a result the magma solidifies around the vent producing steep sided volcanoes called composite or strata volcanoes. They consist both of tephra and lava flows. Mayon volcano in the Philippines and Mt Erebus are examples of strata volcanoes. Very fluid magma of basaltic composition may travel many miles before it solidifies. As a result volcanoes

Mayon Volcano, A Strata Volcano

with gently sloping sides form. These are called shield volcanoes. Mauna Loa and most of the other Hawaiian volcanoes are shield volcanoes. If lava flows from these volcanoes will undergo a change in slope the surface of the lava would become ropy and form a ropy lava. In viscous lava the surface of the lava would break up and become block lava or AA lava. Another type of volcano consists entirely of tephra. It is called a basaltic cinder cone. It is the smallest volcano and lava flows from it will travel for many miles before solidification (due to low viscosity). Many basaltic lava's erupt under the ocean in hot spots or along the mid oceanic ridge. The surface of these lava's cool rapidly due to the water surrounding them, resulting in pillow structures.

Viscosity and composition of magma may also explain the explosives of volcanoes. Viscous, felsic magma does not allow gasses to escape very easily. This results in an increasing pressure in the volcano over a period of time. When the pressure becomes too great for the volcano to withhold violent explosions of the peleean, plinean, vesuvian and vulcanian types occur, peleean being most violent and vulcanian least violent. Nue'e ardentes, lahars and the formation of calderas (collapse of the cone do to circular faults forms a crater or caldera five to 10 kilometers in diameter) are commonly associated with eruptions of these types. Krakatau volcano in Indonesia is a good example of a caldera. Taal volcano in the

Philippines consists of a caldera filled with water, creating a crater lake. Since such violent eruptions require viscous, felsic magma they occur in composite volcanoes.

In very fluid basaltic magma gasses can escape much more easily and frequently. Therefore, large pressure buildups which are associated with violent eruptions do not occur and frequent small explosions do occur. As a result shield volcanoes and basaltic cinder cones have eruptions or are formed by eruptions of the strombolian, hawaiian and fissure types, strombolian being most violent and fissure least . In conclusion, it is evident that composition and viscosity of a magma can go a long way in explaining the origin, shape and processes of a volcano.

Pyroclastics

Pyroclastic rocks are the products of volcanic explosions; that is, they are fragmental pieces of rock, whether they be minerals, crystals or glass, ejected from the vent. Characteristically there are more pyroclastics associated with acid magmas than basic. Acid magmas are more viscous, hence they are reluctant to release gas and this results in high explosiveness. Pyroclastics form by the expansion of gas contained in the parent magma. This may occur when the rising magma comes into contact with ground water. Large amounts of ash from the country rock are associated with eruption when this happens. Pyroclastics may also form when a lava flows into the sea, or even a lake. On such occasions, the products are usually crossbedded breccias that have a foreset dip of approximately 25 degrees. Submarine eruptions also produce pyroclastics, however, the hydrostatic pressure at extreme depths prevents explosions, limiting the production of pyroclastics to shallow depths. Rocks formed from this type of eruption include pumice and hyaloclastites, which is the brecciated product. In general, pyroclastics have been divided into two types:

(1) <u>pyroclastic fall</u> deposits and
(2) <u>pyroclastic flow</u> deposits.

Pyroclastic fall deposits are those which have traveled through the air as some kind of projectile during a volcanic eruption. All ejecta which have traveled through the air are collectively referred to as tephra. Tephra is the main product of many volcanoes, therefore, it is important to be able to recognize the different types of tephra deposits. Tephra is classified according to size and in some instances, shape. A tabulated classification is shown below.

Classification of Pyroclastic Material

> 32 mm	blocks, bombs
>4 mm > 0.32 mm	lapilli, pumice, scoria, etc.
<4 mm > 0,25 mm	Ash
<0.025 mm	fine ash, dust

Bombs are large fragments of rock that form in a characteristic tear drop shape with a twisted tail. They usually contain a central core which is somewhat vesicular. It may be composed of basalt, peridotite, country rock or, rarely, aggregations of crystals, such as olivine. This is covered by an adhering layer of lava which may be glassy in appearance. Bombs may obtain their shape by twisting through the air, but it seems most is derived from the hurling of lithic fragments through molten lava in the throat of the volcano. Dana has suggested that some bombs could have been formed by the rolling movement of the stream front. He names these lava balls. Bread crust bombs form by the contraction of the exterior skin and expansion of the interior; thus forming a cracked surface, exposing the interior. Most bombs vary up to the size of a football (Ollier, 1969) although some very large ones are known. Some have been said to weigh up to 65 tons.

A fragment of rock that is the same size as a bomb, but angular in shape is termed a block. Blocks are made from preexisting rock which probably formed from a previous eruption. Most commonly, they are formed when an explosion tears apart the vent of the volcano, sending angular rock fragments flying through the air. Blocks are definitely not made from molten magma.

Obviously, the large mass of many volcanic bombs and blocks will restrict the distance they are ejected from the vent. Commonly they form heaps of variously sized material adjacent or close to the vent. Much of this material has rolled down the sides of the volcano. These deposits are chaotically arranged with no order to them and are referred to as volcanic agglomerates. They contain lapilli as well as bombs and blocks, and commonly have a dust sized matrix which is post depositional in origin. Their major feature is very poor sorting. This is due to their closeness to the vent. With increasing distance from the vent sorting will be better and average grain size will be smaller, since heavy fragments will not travel as far through the air as smaller fragments will. At some distance from the cone there is sometimes an area which is well sorted but not too fine and suffers a lack of fine ash, meaning it has a large amount of voidspace. The voidspace commonly holds groundwater and this can affect erosion of the area. Agglomerates are sometimes found filling the necks of craters or

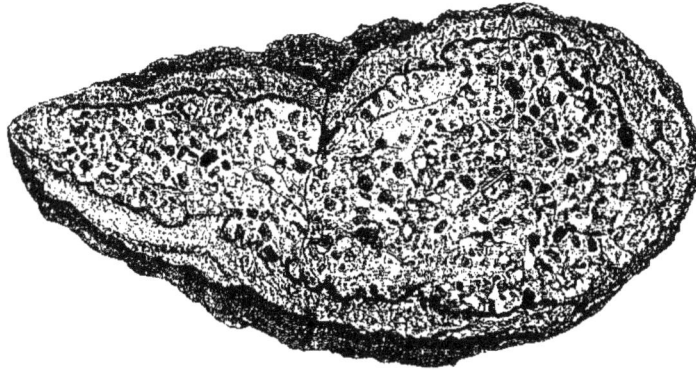

Volcanic Bomb

conduits. These have been observed at the Firth of Fourth (Beckie, 1879). Sometimes agglomerates are found to be well rounded. These are sometimes referred to as volcanic conglomerates and may indicate the presence of water around the volcano. These have also been found associated with non volcanic rocks, such as sedimentary and metamorphic rocks. This may indicate that they were formed from one of the first eruptions in the area. When volcanic activity begins in a new area it begins with an explosion of the crust of the Earth, which may be granite, schist, sandstone, limestone, etc. In areas where volcanic activity stopped after the first eruption all that may be present is a mixture of volcanic and non-volcanic rocks. If volcanism continued, successive layers of volcanic material, representing each eruptive phase, would be piled upon the initial non-volcanic fragments. This has been observed at Haystack Mountain, Montana (Iddings, 1909) where the basal andesitic breccia contains massive amounts of angular gneiss.

Lapilli are commonly associated with bombs and blocks in agglomerates, as well as tuff. Basaltic lapilli which is dark and vesicular are also called scoria or cinders. Acid volcanoes produce highly vesicular lapilli size fragments called pumice. Pumice has such a low density it can often float on water. If the gas bubbles continue to expand it can shatter lapilli into pieces of vesicle walls. These are seen as curved splinters and are known as shards (microscopic). They are common in fine ash and tuff. Lapilli are frequently present in graded beds consisting of bombs and lapilli at the base and ash on the top. These occur due to the falling out of the bombs, followed by lapilli and finally ash, from the air. If strong winds were present at the time of deposition, aeolian crossbedding may occur. Ollier suggests that since the cross bedding is randomly located around some volcanoes it may be due to the blast of the volcanic eruption. At Tower Hill, in Victoria, it has been determined that cross bedding is due to south westerly winds (Marshall, 1967).

Much of the solid material of the cone and the dust which has formed is heated so intensely that it forms ash. Together with dust, ash is the last material to settle after an eruption, sometimes being carried hundreds of kilometers before it's deposited. A good example of this is the recent Mount Saint Helens eruption in which a blanket of ash, up to several inches thick, was deposited as far away as Portland. The characteristic feature of ash; in fact bombs, blocks, and lapilli too, is to follow the lay of the topography, covering hill tops as well as valleys. This is called mantle bedding. Consolidated ash or tuff exhibits this feature. Tuff may be further classified as lithic tuff, if it contains many rock fragments; vitric tuff if most of the fragments are glass; and crystal tuff if well formed crystals are dominant. A general term, tuff breccia or ash fall tuff, is used to describe all of the varieties. Tuff breccia is sometimes bedded in layers representing eruption phases. In some places, such as the Cascade Mountains and the Andes, the thickness of the breccia is 4000 ft and more (Iddings, 1883 & Reiss, 1892). Each bed varies in thickness up to three or four feet. They are usually horizontal or with a dip of about 5 degrees. Some tuffs, especially crystal tuffs, are useful in stratigraphic correlation. Determination of the refractive indices of the phenocrysts and glass in deposits with a known stratigraphic position and comparing them with refractive indices of tuffs from other areas enables correlation. Chemical and mineralogical comparisons may also be done to a limited extent where post depositional processes have not affected the deposits. Where ash overlies dateable materials, such as carbon, accurate ages can be assigned to the deposits. Other information that can be obtained includes: (1) rates of infilling of depositional basins, (2) rate of alluvial fan building; for example, where ash is buried by alluvium, (3) erosion studies; for instance, where there has been no erosion of ash deposited on hillsides, (4) study of sea level changes; for example; at Gisborne, N.S.W. tephra deposits are mantle bedded over old dunes and beach ridges, indicating the paleo-shoreline and role of progradation of the coast, (5) terrace correlation and chronology, (6) archaeology, (7) tectonics. Many other uses of pyroclastic deposits are possible.

The explosion of gas within molten lava can result in the disintegration of the lava into minute pieces of glass and angular fragments collectively called dust. Dust may also form from the massive pulverization of solid material during the explosion. The minuteness of dust particles and the velocity which they are ejected can result in massive dust clouds, finally being deposited hundreds of miles away from the source. The most striking example of this is the eruption of Krakatau in which dust was thrown seventeen miles into the atmosphere and circled the Earth for months before it finally settled, completely disseminated.

The second type of pyroclastic deposits are called pyroclastic flow deposits. The majority of these form when hot fragmenting material made buoyant by hot gas begins to flow as a fluid. This process is called fluidization. It occurs when the hot gasses accompanying the ejecta, together with trapped air and gas being released by the ejecta as it vesiculates, forms an air cushion around each particle preventing it from coming into contact with adjacent particles. Thus, the whole mass behaves like a fluid with low viscosity, enabling it to travel great distances down slopes and, in many instances, up relatively steep slopes. Such clouds of gas can be initiated in several ways. The cloud may just spill over the lip of the volcano and flow downslope as a density current; or an eruption cloud may collapse due to a lack of momentum and heavy load of pyroclastics, forming a pyroclastic flow; or material may pile up on the upper slopes of the volcano and collapse, forming a flow.

There are several types of fluidization flows. General terms used to describe them include ash flows, pumice flows and nue'e ardente or glowing cloud. The term nue'e ardente was derived from the Mt. Pele'e eruption in 1902 which has since become known as the classic example of a pyroclastic flow. Pyroclastic flows have been classified into dense, intermediate, and vesicular flows; and ash flow to vesicular types. A more detailed classification is shown in the table below:

Pyroclastic Flow Classification

Dense	Nue'e Ardente
	    • Pelee' type (flow from side of dome)
	       Merapi type (flow from collapsing dome)
	       Lakurajiima type (flow from open crater)
		0.001 - 0.3 Km^3

Intermediate	Intermediate	0.05 - 1 Km^3

Vesicular	Ash Flow
	    • St. Vincent (Vert. eruption from crater)
	       Krakotoa (as St. Vincent but > magnitude)
	       Valley of Ten Thousand Smokes (magma discharge through fissures)
		0.1 - 90 Km^3

The term used to describe the type of deposit formed this type of pyroclastic flow is ignimbrite. Terms such as ash flow deposit, welded tuff, tuff flow deposit and nue'e ardente deposit have also been used. Ignimbrite is a general term and it is specifically used to describe the rock unit. Silicic ignimbrites are most abundant, although ryholite to basaltic ignimbrites also occur. The length of the deposit and area of deposition depends upon the topography. Ignimbrites are usually deposited in low lying areas, such as valleys, since the density of the flow does not permit it to stop on high ground. This is in contrast to pyroclastic fall deposits, which are mantle bedded.

Ignimbrite deposits generally consist of three layers. The basal layer, or sillar layer, generally has an absence of large fragments which are probably expelled by frictional forces. It is of a finer grain size with small glass and pumice fragments and is generally a white or gray color. The middle layer is often welded and is very poorly sorted. It often shows a reverse grading of pumice blocks and normal grading of large lithic fragments. It varies in color from a brown to deep black. The middle layer merges into the top layer which consists of unwelded ash tuff.

Another type of deposit which is not strictly a flow deposit but which is commonly associated with and mistaken for one is a surge deposit. Base surge deposits, also called Maar deposits, are due to the horizontal blast from an explosion transporting material. They are usually found on flat ground, are low density deposits and form similar structures to those found in a river bed (thin bedded with localized dome shaped crossbeds). Ground surge deposits are similar to base surge deposits in origin. They mainly form from central crater type volcanoes where gravity forces as well as horizontal blast forces assist their flow. They are less than 10m thick and can be composed of juvenile magma, lithic fragments, and crystals in any combination or proportion.

Lahars, or mudflows of volcanic material, are very common in high rainfall areas. They form from fresh ejecta, which is still hot, mixed with water or cold ejecta which is mixed with rain water or surface water. They are very mobile, poorly bedded, with occasional layers of crossbeds and have poor sorting. They characteristically form hillocks, ranging in size from a few meters to tens of meters, with a core of boulders. They sometimes have fine basal regions and can be indistinguishable from ignimbrite deposits. Generally, flows with greater than 10% of water content are called mud flows and flows with less than 10% are called debris flows. Debris flows consist of highly concentrated, high density material.

Avalanche deposits can also resemble ignimbrites. They usually form from very viscous, acidic lavas which are almost solid when ejected. They crumble, break up, and collapse downslope (hot rock avalanche if they are vesicular). They resemble ignimbrites and are best distinguished by their overall geometry.

Two types of pyroclastic deposits are recognized, pyroclastic flow deposits and pyroclastic fall deposits. Pyroclastic fall deposits usually have good sorting and cover vast areas. Pyroclastic flow deposits are commonly poorly sorted and sometimes welded. It is important for the geologist to be able to recognize these deposits because of their economic significance, not only in terms of mineralization but also in environmental interpretation.

Plate Tectonics

By Rob Kanen

Plate Tectonics is one of the most important geophysical/structural geology subjects today. To determine the cause of the movement of the plates is the most studied problem. The first evidence for plate movement was, of course, found by Wegener in 1925. This was a result of a comparison of the continental edges of South America and South Africa. It was not until the 1950's, however, that Carey (1954) found the remarkably good fit between the continents using a modelled globe. Wegener's evidence was primarily geological and paleo-climatological.

The model of the Earth developed by the seismologists, at this time, was a liquid iron core surrounded by a solid mantle with no convection movements. When Elsasser and Bullard (1965) developed their geomagnetic field theory, postulating that there are convective motions in the fluid iron core, there was no real objection by the seismologists since the core did not transmit s-waves, indicating it is a classical fluid. It was not until the development of paleomagnetism that there was new evidence for continental drift, then later on, geophysical measurements of the ocean floor swept away most of the doubts geophysicists had about continental drift. This now constitutes part of the subject called plate tectonics.

Many theories on the mechanism for plate movement have been developed. The most popular and widely held view is that convection currents below the lithospheric plates, in the mantle, are responsible for their movement. This involves hot spots and subduction zones. The most radical view was that that developed by Carey (1954), Heezen (1959) and others , that the Earth is expanding causing the continents to break up and form plates.

The Plume Hypothesis

Morgan (1971, 1972) advocates the idea of mantle plumes to explain continental drift. Briefly, he advocates deep mantle convection in which narrow plumes of deep material rises and spreads out laterally in the asthenosphere. This convective movement causes stresses on the bottoms of the lithospheric plates, causing them to move. He suggests that "hot spots", areas of upwelling visible in the Earth's surface, provide the motive force for continental drift. It is based on three facts: (1) Most of the hot spots are near a ridge and a hot spot is near each of the triple ridge junctions; (2) the gravity and regional high topography suggests that more than just surface volcanism is involved at each hot spot; and (3) neither rises nor trenches appear capable of driving the plates, implying that asthenospheric currents acting on the plate bottoms must exist. He bases his theory on data and observations made worldwide. This explanation is convincing as his observations are simple and sound.

Runcorn (1980) discounts Morgan's reasoning because it is based on the analogy with plumes in the asthenosphere; that a plume maintains a small horizontal width as it rises to a very great height. The reason this occurs in the atmosphere, according to Runcorn, is because the inertia term in the Navier-Stokes equation is much greater than the viscous force. In the mantle, the reverse is true.

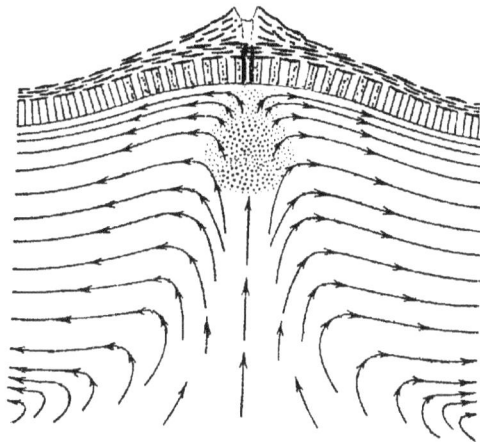

Mantle Plume

Gravity Sliding

Before the evidence for convection became known, geophysicists tried to explain plate movement as do to their own inherent properties ie. gravity force. Hales (1969) suggested that the plates were moving away from the mid-oceanic ridges. Isac's and Molnar(1969), after the discovery that the plates were sinking into the asthenosphere along the trenches, suggested that since the overhanging part of the plate was colder than the surrounding mantle, it would also be denser, thus, a downward gravity force might cause the horizontal movement of the plate. Runcorn (1974) showed that by using magnitude calculations, it is possible a sufficient force would be produced. Therefore, this theory cannot be rejected on the grounds of magnitude, but should be rejected because it is impossible to explain how the process began and it does not enable any understanding of plate movements. Morgan (1972) rejected this theory, but on the grounds that small trench bounded plates, such as the Cocos, do not move faster than the larger Pacific Plate as would be expected.

Convection

In contrast to the Plume theory of convection (Morgan, 1972), Runcorn (1980) promotes the theory of large-scale convection. He believes the only way the continents and plates could move in the regular way they have for the past hundred million years is by convection ion a large scale cell structure. This convection pattern changed with time from a one cell, to a two cell, to a three cell and then to a four cell pattern; thereby, explaining the breakup of Gondwanaland and Laurasia (Pangea) in only the last 150 million years. Evidence for this is sparse, however, one reason for large scale convection is that the Earth formed by accretion with the heavy elements, such as iron, sinking to the center forming a core (Urey, 1951); thus, the core may be continually growing. Runcorn (1980) says it is expected there would be a greater geothermal gradient in the lithosphere above the rising convection currents, thus, it would be possible to find the size and location of the currents. However, the time constant for the lithosphere 100m thick is about 100 million years in which time a plate would have moved a considerable distance relative to the origin of a heat source.

Runcorn (1980) uses the shape of the geoid to support this theory. He says, "if a planet departs from hydrostatic equilibrium on a large scale then there are only two possible explanations. Either, density anomalies were acquired by the planet in its early history or the distortion of the planet is being produced dynamically". In otherwards, "...the geoid is the primary evidence for convection patterns in the mantle". Jeffrey's (1975) has discovered the Earth's gravitational field departs from the accepted hydrostatic equilibrium model. However, the shape of the geoid and its relationship remain open to interpretation.

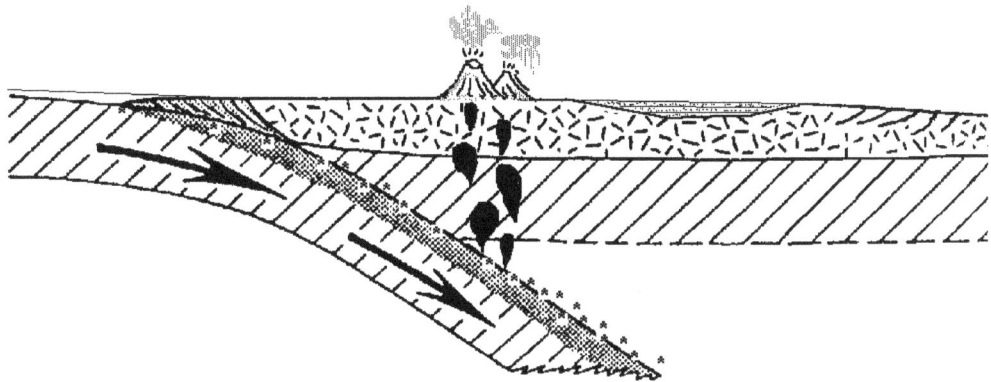

Subduction

The Expanding Earth

Numerous authors, such as Egyed (1957), Cox and Doell (1961), Ward (1963), Creer (1965), Heezen (1960) and especially Carey (1954, 1970) have supported the theory that continents have moved apart because of an expanding Earth. Carey based his theory on geologic and tectonic observations while most other authors have used paleomagnetic data to supplement his initial theory. Carey (1970) proposes the Earth is made up of eight first order polygons, analogues to the lithospheric plates, with accretion occurring on all sides of each polygon. He says sea floor spreading supports his argument, that new crust must be forming between continents for expansion to occur. Thus, each of them has increased greatly in area, irrespective of how much or how little swelling of crust along trenches occurred. Thus, this means the Earth has increased in total surface area by a large amount. Based on the area of oceanic crust on each polygon, Carey has calculated the amount of expansion, which has taken place, is 76%. This equivalent to a 33% increase in radius.

As further evidence, Carey suggests the following: if you stand on any polygon, it has moved away from every other polygon and if you face about, the distance to each polygon has also increased. This is assuming the Earth consisted almost entirely of continental crust with oceanic crust only being produced in the Mesozoic-Phanerozoic. There is evidence that oceanic crust must have been present before this time, for instance, along Lower Paleozoic accreted continental margins and in Precambian greenstones.

That oceanic area has increased is consistent with Egyed's observations that each polygon shows progressively less marine transgression through geologic time. Since oceanic crust can have twice as much water as continental crust Carey (1970), the theory that the Earth's surface has increased with time by progressive increase in the area of the ocean basins is supported.

Island arcs appear to be supportive of a subduction type environment in which case convective movement on a non-expanding Earth would be the obvious mechanism for plate movement. However, Carey (1970) denies that island arcs are the result of volcanism along a zone where oceanic crust is being subducted under continental plate. Instead, because all island arcs are bowed eastward, he suggests they are tensional features. To illustrate this, Carey used an analogy with a glacier. On the western side is a dilation zone of new oceanic crust with high heat flux and repeated horsts and grabens, in contrast with the other side which is passive and quiet with little disturbed sediments. Dilation rifts occur in similar fashion at the head of a glacier and the graben arc. Thus, he says, " trenches are dilation rifts". This theory is in stark contrast to calculations made by various authors, such as Morgan (1972) and Runcorn (1980), who state the plates are actually colliding, moving in opposite directions. This data is widely accepted as the norm today.

Carey (1970) makes a bold suggestion when he says his data indicates expansion must be occurring at the rapid rate of 8mm a year and almost all of it occurring since the Late

Paleozoic-Mesozoic. Almost universally, other authors (Dooley, 1973; Creer, 1965; Egyed, 1960) have through their calculations based on paleomagnetic data suggest this is impossible. They have in turn, however, stated that the Earth could expand at a slower rate, upto 0.5mm a year, for long periods.

These authors, Creer (1965) in particular, suggest the Earth may have been expanding all its life at a slow rate. Therefore, at the time Pangea began to breakup, the Earth's radius would have been similar to its present radius. Numerous models have been constructed (Dooley, Creer) to illustrate how well the continents fit together to form Pangea. All agree that the best is obtained at present. This destroys Carey's model, which assumes to have a radius 76% of the present radius at the time of Pangea breakup. This does not destroy the suggestion that the Earth is expanding at a slower rate.

Creer (1965) says, "I think expansion should be regarded as something which may been gently, but persistently, occurring in the background. There may be little obvious geological evidence of expansion, most of this could easily have been obscured by more vicious and rapid processes, such as continental drift and orogeny." He goes on to say that to obtain a satisfactory explanation of expansion we may well have to wait until the origin of the universe has been successfully deciphered.

Conclusion

The best theory contains only minor holes and explains the mechanism of plate tectonics in a simple, clear and distinct way. Convective plume theory, developed by Le Pichon (1968), Morgan (1968), Runcorn (1980) and others has three major flaws: (1) plate boundaries are not distinct; (2) the condition that each plate having its own accretion and consumption boundary, as for the case for the African Plate, is violated; and (3) if the plates are rigid, as assumed, deformation should have occurred in bottle necks where part of a plate margin was subducted and the rest was not. Of course, the presence of island arcs, subduction zones, hot spots and basalt relationships support the convective-plume theory.

Introduction to Geology

The expansion theory of Cary has major flaws in it, among others, these are: (1) that the Earth was assumed to consist entirely of continental sialic crust; and (2) that a rapid expansion at a rate of 8mm/year had to occur in the last 200my; and (3) that the Earth had radius 76% of its present radius when Pangea broke up.

The slow-expanding Earth theory of Creer (1965) and others is more plausible but lacks evidence. It does not suggest why the Earth would expand, why continental drift began so late in the Earth's history or where the energy source for expansion is derived from. The conclusion is that the convective-plume theory is the most plausible, based on evidence available.

Appendix 1: Geologic Time Scale

The following four charts represent geologic time from the time of the oldest recoreded rocks to the present. The oldest dated rocks are from the Archean, approximately 4.3 billion years old. The Earth is estimated to be 4.5 billion years old. The geologic time scale is a result of over a hundred years of scientific study into the age of our planet, particularly the rocks, minerals and fossils that comprise the surface and near surface. During the first fifty or so years of study, ages were determined by studying the relative relationships of different rock units and the fossils they contained, a subject known as startigraphy. Later on, radiometric dating of isotopes, using mass spectrometers, became the most reliable method of age dating rocks. This is the most widespread method used today for accurately dating rocks. The field method of stratigraphic correlation is still widely used on regional scales, particularly for geologic mapping, and when radiometric dating is not available.

EON	ERA	PERIOD	EPOCH	STAGE	Ma	EVENT
PHANEROZOIC	CENOZOIC TERTIARY	QUATERNARY	PLEISTOCENE	Calabrian	1.78	
		NEOGENE	PLIOCENE	Piacenzian	3.56	
				Zanclian	5.32	
			MIOCENE	Messinian	7.12	First appearance of hominids
				Tortonian	11.2	
				Serravallian	14.8	
				Langhian	16.4	
				Burdigalian	20.52	Red Sea Opens
				Aqitanian	23.8	
		PALEOGENE	OLIGOCENE	Chattian	28.5	
				Rupelian	33.7	
			EOCENE	Priabonian	37	
				Bartonian	41.3	
				Lutetian	49	
				Ypresian	54.8	Australia and Antarctica begin seperating
			PALEOCENE	Thanetian	57.9	First appearance of primates
				Selandian	60.9	
				Danian	65	
	MESOZOIC	CRETACEOUS	LATE	Maastrichtian	73	Extinction of ammonites, dinosaurs, globotrucanid foraminifera
				Campanian	83	Tasman Sea opens
				Santonian	87	
				Coniachian	89	
				Turonian	91	
				Cenomanian	97.5	First appearance of marsupials, placentals
			EARLY	Albian	108	
				Aptian	115	
				Barremian	123	First appearance of diatoms
				Hauterivian	130	First appearance of angiosperms
				Valanginian	135	Labrador Sea opens
				Berriasian	141	
		JURASSIC	LATE	Tithonian	146	South Atlantic opens First appearance of birds
				Kimmeridgian	151	India, Madagascar and Antarctica seperate
				Oxfordian	159	
			MIDDLE	Callovian	165	
				Bathonian	173	First appearance of globireginacea foraminifera.
				Bajocian	180	Gondwana and Laurasia seperate
				Bernasian	184	

EON	ERA	PERIOD	EPOCH	STAGE	Ma	EVENT
P H A N E R O Z O I C	M E S O Z O I C	JURASSIC	EARLY	Toarcian	190	
				Pliensbachian	195	
				Sinemurian	202	
				Hettangian	205	
		TRIASSIC	LATE	Rhaetian	207	Extinction of conodonts
				Norian	220	Diversification of reptiles
				Carnian	230	
			MIDDLE	Ladinian	235	
				Anisian	241	First appearance of dinosaurs, hexacorals Last appearance of Rugose corals
			EARLY	Spathian	244	
				Nammalian	247	
				Griesbachian	251	Extinction of trilobites, tabulate corals, orthid brachiopods
	P A L E O Z O I C	PERMIAN	LATE	Tatarian	256	
				Kazanian		Proto-Atlantic closes
				Ufimian	267	
			EARLY	Kungurian	274	
				Artinskian	274	
				Sakmarian	285	
				Asselian	293	
		CARBO-NIFEROUS	STEPHANIAN	Gzelian	298	First appearance of winged insects
				Kasimovian	300	
			WESTPHALIAN	Moscovian	305	First appearance of pleisiosaurs
				Bashkerian	311.5	
			NAMURIAN	Serpukhovian	314	
				Brigantian	325	
			VISEAN	Asbian	328.5	
				Holkerian	330.5	
					336	

EON	ERA	PERIOD	EPOCH	STAGE	Ma	EVENT
PHANEROZOIC	PALEOZOIC	CARBON-IFEROUS	VISEAN	Arundian	341.5	
				Chadian	344	
			TOURNASIAN	Ivorian	348	
				Hastarian	354	Extinction of graptolites
		DEVONIAN	LATE	Famennian	384.5	First appearance of amphibians
				Frasnian	369	
			MIDDLE	Givetian	378	
				Eifelian	384	
			EARLY	Emsian	399.5	First appearance of ammonites
				Pragian	404.5	First appearance of land plants
				Lochkovian	410	
		SILURIAN	PRIDOLI		414	
			LUDLOW	Ludfordian	417	
				Gorstian	420	
			WENLOCK	Homerian	422.5	
				Sheinwoodian	425	
			LLANDOVERY	Telychian	428	
				Aeronian	431	
				Rhuddanian	434	
		ORDOVICIAN	ASHGILL	Bolindian	443	
			CARADOC	Eastonian	454.5	
				Gisbornian	459	
			LLANVIRN	Darriwelian	467	

Introduction to Geology

EON	ERA	PERIOD	EPOCH	STAGE	Ma	EVENT
P H A N E R O Z O I C	P A L E O Z O I C	ORDOVICIAN	ARENIG	Yapeenian		
					471	
				Castlemainian		
					477	
				Chewtonian		
					481	
				Bendigonian		
					484.5	
			TREMADOC	Lancefieldian / Warendan	486	
					490	
		CAMBRIAN	LATE	Pantonian-Datsonian	492.5	
				Iverian	496	
				Idamean-Mindyalian	498.5	
			MIDDLE	Boomerangain	500	First appearance of echinoids, bryozoans
				Undilian	503	
				Late Templetonian-Floran	506	
				Ordian-Early Templetonian	509	
			EARLY	Atdabanian-Toyonian	528	
				Tommotian	534	
				Nemakit-Daldynian		First appearance of graptolites, fish (vertebrates)
					545	First appearance of exoskeletons
P R O T E R O Z O I C	SIN-EAN	VENDIAN	EDIACARAN		590	
			VARANGIAN		610	
		STURTIAN			800	
	RIP-HEAN	KARATAU			1050	
		YURMATIN			1350	First appearance of organisms with a nucleus (eukaryotes)
		BURZYAN			1650	
		ANIMIKEAN			2200	First appearance of red beds. Last appearance of banded iron stones
		HURONIAN			2450	
A R C H E A N		RANDIAN			2800	First appearance of stromatolites, microorganisms
		SWAZIAN			3500	Oldest sedimentary rocks
		ISUAN			3800	Oldest dated rocks
		HADEAN	EARLY IMBRIAN		3850	
			NECTARIAN		3950	
			BASIN GROUPS 1-9		4150	
			CRYPTIC		4560	

(2500 appears near the 2450 Ma line; 4000 appears near the 3950 Ma line)

Appendix 2: Periodic Table

The periodic table is a tabular listing of the chemical elements. The elements are listed according to their atomic number and properties, which reflect their periodic nature. The modern periodic table is attributed to the work of the Russian chemist Mendelev in 1869. Mendelev left gaps in his original periodic table that were later filled in by other chemists with newly discovered elements. This predictable and periodic nature of the table makes it most usefull as an indispensible reference for all scientists.

1 IA	2 IIA	3 IIIB	4 IVB	5 VB	6 VIB	7 VIIB	8	9 –VIIIB–	10	11 IB	12 IIB	13 IIIA	14 IVA	15 VA	16 VIA	17 VIIA	18 VIIIA
1 H 1.008																	2 He 4.00260
3 Li 6.941	4 Be 9.01218											5 B 10.81	6 C 12.011	7 N 14.0067	8 O 15.9994	9 F 19.9984	10 Ne 20.179
11 Na 22.9898	12 Mg 24.305											13 Al 26.9815	14 Si 28.086	15 P 30.9738	16 S 32.06	17 Cl 35.453	18 Ar 39.948
19 K 39.102	20 Ca 40.08	21 Sc 44.9559	22 Ti 47.90	23 V 50.9414	24 Cr 52.996	25 Mn 54.9380	26 Fe 55.847	27 Co 58.9332	28 Ni 58.71	29 Cu 63.546	30 Zn 65.37	31 Ga 69.72	32 Ge 72.59	33 As 74.9216	34 Se 78.96	35 Br 79.904	36 Kr 83.80
37 Rb 85.4678	38 Sr 87.62	39 Y 88.9059	40 Zr 91.22	41 Nb 92.9064	42 Mo 95.94	43 Tc (98.9062)	44 Ru 101.07	45 Rh 102.9055	46 Pd 106.4	47 Ag 107.868	48 Cd 112.40	49 In 114.82	50 Sn 118.69	51 Sb 121.75	52 Te 127.60	53 I 126.9045	54 Xe 131.30
55 Cs 132.9055	56 Ba 137.34	57 *La 138.9055	72 Hf 178.49	73 Ta 180.9479	74 W 183.85	75 Re 186.2	76 Os 190.2	77 Ir 190.22	78 Pt 195.09	79 Au 197.9665	80 Hg 200.59	81 Tl 204.37	82 Pb 207.2	83 Bi 209.9806	84 Po (210)	85 At (210)	86 Rn (222)
87 Fr (223)	88 Ra (226)	89 #Ac (227)	104 Rf (257)	105 Db (260)	106 Sg (263)	107 Bh (262)	108 Hs (265)	109 Mt (266)	110 Ds (272)								

*	58 Ce 140.12	59 Pr 140.9077	60 Nd 144.24	61 Pm (145)	62 Sm 150.4	63 Eu 151.96	64 Gd 157.25	65 Tb 158.9254	66 Dy 162.50	67 Ho 164.9303	68 Er 167.26	69 Tm 168.9342	70 Yb 173.04	71 Lu 174.97
#	90 Th 232.0381	91 Pa (231.0359)	92 U (238.029)	93 Np (237.0482)	94 Pu (244)	95 Am (243)	96 Cm (247)	97 Bk (247)	98 Cf (251)	99 Es (252)	100 Fm (257)	101 Md (258)	102 No (259)	103 Lr (260)

Alkali Metals

Alkali Earth Metals

Transition Metals

Halogens

Noble Gases

Rare Earth Metals

Other Metals

Cu–Solid   Hg–Liquid   O–Gas   Tc–Synthetic

References

The Chapter on "Minerals and Rocks" is from
Longwell, C. R., Knopf, A., Flint, R. F., 1939, "Textbook of Geology", Vol. 1, N. Y., John Wiley and Sons , New York.

The Chapters on "The Principles of Petrology" and "Rocks and Their Composition" are from
Tyrrell, G. W., 1929, "The Principles Of Petrology: An Introduction to the Science of Rocks", 2nd Edition, New York, 349p

Rock Classification and Petrology:

Cox, K.G., Bell, J.D. and Pankhurst, R.J., 1979, "The Interpretation of Igneous Rocks" Allen and Unwin, London, 450p. - Norm calculations and indices.

Le Maitre, R.W.(ed), 1989, "A Classification of Igneous Rocks and Glossary of Terms". Blackwell Scientific Publications, Oxford, UK. - Igneous plots and systematics.

McBirney, A, 1993, "Igneous Petrology". Jones and Bartlett. - Norm calculations and indices.

Myashiro, A.,1973. "Metamorphism and Metamorphic Belts". Allen and Unwin, London,492p.

Pettijohn, E.J., 1975, "Sedimentary Rocks". Harper and Row. - Arenites and Wackes.

Selley, R.C, 1981, "An Introduction to Sedimentology". Academic Press Inc, London. - Sandstone plots

Granites

Carmichael, Turner and Veerhoogen, "Igneous Petrology", McGraw Hill, 1974
Chappell, B.J. and White, A.J.R., 1974, "Two Contrasting Granite Types". Pac. Geol., v8, pp.173-174.
Flood, R.H. and Shaw, S.E., 1975,"A Cordierite bearing granite suite from The New England Batholith". Contr. Mineral. and Petrol., v52, pp.157-164.

Gilluly, I, "The Tectonic Evolution of the United States", Q.J. Geol. Soc. London, v.119, 1963

Hine, R., Williams, I.S., Chappell, B.W. and White, A.J.R., 1978, "Contrasts Between I- and S-Type granitoids of the Kosciusko Batholith", J. Geol. Soc. Aust., V25, pp.219-234.

O'Neill, J.R. and Chappell, B.W., 1977, "Oxygen and Hydrogen Isotope Relations In the Berridale Batholith", J. Geol. Soc. Lond., V.133, pp.559-571.

Pitcher, W.S., "The Nature, Ascent and Emplacement of Granitic Magmas", J. Geol. Soc. of London, v.136, pp. 627-662, 1979

Read, H.H.,"The Granite Controversy", Murby, 1957.

White, A.J.R. and Chappell, B.W., "Ultrametamorphism and Granite Genesis", Tectonophysics, V.43, pp. 7-22.

Plate Tectonics

Carey, S.W., 1970, "Australia, New Guinea and Melanes in the Current Revolution in Concepts of the Evolution of the Earth". Search, V.1, No.5, pp.178-189.

---------------, 1976, "The Expanding Earth". Elsevier, Amsterdam, 488p.

Chudinov, Y,V., 1977, " Expansion of the Earth as alternative to the New Global Tectonics". Geotectonics, v.10, pp240-250.

Cox, A. and Doell, V.R., 1961. Nature, v.189, p45.

Creer, K.M., 1965, "An Expanding Earth" Nature, v205,, no.4971, pp.539-544.

Dooley, J.C., 1973, "Is the Earth Expanding?". Search, v4, no.1-2, pp.9-15

Egyed, L., 1961, " Nature, v190, no.1097.

Elasser and Bullard, 1965

Hales, 1969

Heezen, B.C., 1960, Sc. Amer., Oct., v.3.

Le Pichon, X., "1968, "Sea Floor Spreading and Continental Drift". J. Geophy., pp.3661-3697.

Molnar, 1969

Morgan, W.J., 1971,

Morgan, W.J., 1972, "Plate Motions and Deep Mantle Convection". Geol. Soc. Amer., Mem 132, pp.7-22.

Runcorn, S.K., 1980, "Mechanism of Plate Tectonics: etc.". Tectonophys., v.63, pp. 297-307.

Ward, M.A., 1963, Geophy. J., v.8, p.217

Wegener, A., 1925, " Origin of the Continents and Oceans". Gos izdat.

Volcanoes etc.

Fielder, G & Wilson, L "Volcanoes of the Earth, Moon and Mars: Pg 49-56, 1975

Iddings "Igneous Rocks" 1909

Holmes "Principles of Physical Geology" Ed. 3

Ollier, C "Volcanoes" Vol. 6 1969

Strahler, A, N "Principles of Physical Geology" 1977

Picture Credits

Diagrams in the Chapters "Granites", "Volcanoes" and "Plate Tectonics": Gibson Consulting

Geologic Time Scale and Periodic Table: Rob Kanen

All other photos and illustrations: Rob Kanen

Introduction to Geology

www.ingramcontent.com/pod-product-compliance
Lightning Source LLC
Chambersburg PA
CBHW080542220326
41599CB00032B/6336